The [barcode]

C000057261

Max Adams is the auth[...]
Collingwood: Nelson's Ow[...]
an archaeologist Max turned to another passion and became a
woodsman. He has since combined teaching with travelling,
writing and broadcasting. He lives and works in the North-east
of England with his son, Jack.

'Historians and biographers like to emphasise the galvanic effect
of science on culture during what we have come to call the
Romantic age. Max Adams has undertaken something new . . .
he has done it dazzlingly . . . Martin's own 90mph ride with his
friend Isambard Kingdom Brunel, along six miles of the engineer's
new Great Western Railway, can hardly have been more
exhilarating than this tearaway voyage through stirring times'
Guardian

'If Shelley was the 'prophet of Prometheanism', then the
romantic painter John Martin was its high priest . . . Writing of
an age before rigid disciplinary boundaries, Adams illuminates
the links between a generation of artistic and scientific visionaries'
Independent on Sunday

'A panoramic account of early-nineteenth-century artistic,
scientific and political radicalism . . . providing an engaging
perspective on the era in which [John Martin] lived'
The Times

'An intriguing look at the impact on Britain of the first industrial
revolution . . . what makes this book so fascinating is how such
an ostensibly ordinary family came to reflect all the political,
philosophical, scientific and artistic changes of the age.
Wonderfully eclectic in its content'　　　*Good Book Guide*

ALSO BY MAX ADAMS

Admiral Collingwood: Nelson's Own Hero

The Prometheans

John Martin and the Generation
that Stole the Future

MAX ADAMS

Quercus

This paperback edition published by Quercus in 2010
Originally published as *The Firebringers* in 2009 by

Quercus
21 Bloomsbury Square,
London
WC1A 2NS

A CIP catalogue record for this book is available
from the British Library

ISBN 978 1 84916 173 2

Typography by Gwyn Lewis

Printed and bound in Great Britain by Clays Ltd, St Ives Plc

10 9 8 7 6 5 4 3 2 1

FOR WARWICK

Emancipate yourselves from mental slavery
None but ourselves can free our minds

Bob Marley, 'Redemption Song'

Contents

Acknowledgements

MY THANKS MUST GO FIRSTLY TO the judges and sponsors of the Elizabeth Longford Award, whose generous support helped me with the research for this book at a crucial time. Flora Fraser has been a kind and encouraging advocate of the *Firebringers* project and I am most grateful for her advice.

I owe a debt of gratitude to the innumerable scholars and authors whose work I have leant upon, especially Thomas Balston and William Feaver for their detailed and inspired biographies of the Martins. I would like to thank Henry Swaddle, a Martin scholar and friend, whose 'World of John Martin' website (http://www.wojm.org.uk) is the outstanding public source for the Martins. Other students, with whom I have been privileged to share my enthusiasms, have contributed much to my thinking on the Private; none more so than Keith Tulip, who tracked down an important portrait of Karl Marx as Prometheus and offered many provoking thoughts.

Many people helped with my research. I wish to thank the staff of the Literary and Philosophical Society of Newcastle upon Tyne, the Royal Bethlehem Hospital, the British Library, the Ashmolean Museum print room, Newcastle City Library and the Laing Art Gallery for their kind assistance.

For contributing important notes or answers to annoying questions I would like to acknowledge the help of Robert Rennison, John

Gath, Mick Manning, Rod Williams, Anna Flowers, Jim Rees and Mari Takayangi, archivist of the Parliamentary Archives at the Houses of Parliament.

I am indebted to those kind friends who read and improved the manuscript, especially Sally Kidman, Dr Chris Cumberpatch, and Professor John V. Pickstone. Thanks are also due to Dan Elliott and to my father, Warwick Adams, for their support and critical input at all stages of the *Firebringers* project.

List of illustrations

PAGE 5

ABOVE *London's Overthrow* by Jonathan Martin, drawing/Bethlehem Royal Hospital
LEFT *Sadak in Search of the Waters of Oblivion* by John Martin, oil on canvas/Southampton City Art Gallery, Hampshire, UK/Bridgeman Art Library

PAGE 6

TOP *Sir Marc Isambard Brunel* by James Northcote, oil on canvas 1812–13/National Portrait Gallery/Bridgeman Art Library
BOTTOM LEFT Davy Lamps, 1815/Science Museum/SSPL
BOTTOM RIGHT *Sir Humphry Davy* by John Linnell, oil on canvas/Private Collection/Bridgeman Art Library

PAGE 7

TOP LEFT Cooke and Wheatstone, five-needle telegraph, 1837/Science Museum/SSPL
TOP RIGHT Challenge board for Bramah padlock, 1801/Science Museum/SSPL
BOTTOM LEFT Charles Babbage's Difference Engine No. 1, 1824–32/Science Museum/SSPL
BOTTOM RIGHT Charles Babbage engraving, 1832 English School/Private Collection/Bridgeman Art Library

PAGE 8

TOP Joseph Mallord William Turner, *Self-portrait*, 1798, oil on canvas/National Gallery, London/Bridgeman Art Library
BOTTOM Joseph Mallord William Turner, *Rain Steam and Speed – The Great Western Railway*, painted before 1844, oil on canvas/National Gallery, London/Bridgeman Art Library

Forethought

JULY 18TH 1822. The half-mutilated corpse of a drowned poet is washed up on the beach at Viareggio in Italy. In one pocket of his ragged salt-crusted jacket is a volume of poetry by Keats. In the other, either Sophocles or Aeschylus, no one is quite sure. By the time a funeral can be organized the body of Percy Bysshe Shelley, dead three weeks before his thirtieth birthday, is an unrecognizable bundle of disarticulated flesh and bone. Tattered, sodden black rags are all that remain of his clothes. His face is missing.

Shelley's friends Edward John Trelawney, Leigh Hunt and Lord Byron arrange for a funeral pyre to be built on the beach in imitation of an ancient Greek rite and stand solemnly watching as the poet's mortal remains are burned to ashes, scarcely able to believe that this man, of such rare physical and intellectual beauty, is gone for ever.

Shelley was a polemical lyricist, a restless libertine, a revolutionary, an atheist, England's great white hope in a contradictory age of aspiration, cynicism, tolerance and repression. He was the man whose words might have altered the course of British history in the nineteenth century had they been publishable: less obviously seditious or inflammatory. He was the man who, if he had raised a standard in his own country, could have led a Jacobin uprising every bit as bloody as the Parisian Terrors, though to be sure he was no Marat, no Robespierre. He was the man who wrote a manifesto for a new world in the ancient ruins of the Caracalla Baths in Rome.

In 1822 England needed Shelley. While he had been in Italy cavalry

1

sabres had cut down peaceful protesters on the streets of Manchester. Political activists languished in gaol. Shelley's 'old, mad, blind, despised' King George III had died and been replaced by his son: a corpulent libertine, the 'dregs of his dull race'. An ancient, feudal landowning elite, who 'leech-like to their fainting country cling' were subsidized by Corn Laws to keep bread out of the mouths of the poor, their ranks swelled by a hundred thousand and more demobilized servicemen. The press was gagged by stamp duties and Treason laws. *Habeas corpus* had been suspended. The reform movement, for a generation emasculated by war and repressive legislation, desperately sought the inspiration of a popular leader, a rallying voice.

It is true that something of an opposition existed in Britain. That old rabble-rousing dog William Cobbett snapped at the heels of the Ministry. Arthur Thistlewood and his land-sharing conspirators plotted incompetently to murder the Cabinet. The Hunts' *Examiner* chided and carped. Lord Byron took Their Lordships to task in the Upper House while Henry Brougham nagged the Commons. Cruikshank, Gillray, Martin, even Turner painted words of warning on the wall for those perceptive enough to read the message. But none of these men was Shelley.

A broad coalition of artists, writers, scientists, engineers and educators, many of them born between the American and French revolutions, might have marched under Shelley's banner if he had called them to arms. He more than any of them articulated what it was to be part of this generation of firebringers, the first perhaps in history, who sought to change the world for the better by liberating all humans from the tyrannies of oppression and faith.

Shelley was, perhaps, the only one of them explicitly to recognize himself as the prophet of the cult of Prometheus (the Greek Titan whose name in English means 'forethought'). This unacknowledged sect, which celebrated the original theft of the secret of fire, had its political origins in the republican movement of the English Civil War and the poetry of Milton. Its intellectual inspiration came from the Germany of Kant, Goethe and Hegel.

The high priest of the French Enlightenment, Jean-Jacques Rousseau, recognized the Promethean myth as a liberating emblem for radicals, as did American men of action like Thomas Jefferson and Benjamin Franklin. In England the cult attracted radical visionaries: William Blake, Tom Paine and William Godwin, Shelley's own father-in-law.

Among Shelley's contemporaries in England, Prometheanism came to its full flowering. The Titan's theft of fire from Zeus as a disinterested gift to humanity and his subsequent punishment chained to a rock, inspired artists to reinvent colour and light; scientists to reveal the secrets of elemental chemistry and electricity; engineers to forge steam-belching workhorses, and radicals to prise the lid from Pandora's jar and peer inside. Free the slaves, emancipate the Catholics, educate the masses and above all reform Parliament: these were merely their secular demands.

There were anti-Prometheans too. Mary Wollstonecraft, who married William Godwin, was one. Her daughter, Shelley's wife, was another in whose *Frankenstein, or the Modern Prometheus* the sin of hubris is punished like Prometheus himself: with eternal torment. A broad conservative coalition that included the monarchy and Tory landowners, reactionaries like Edmund Burke and even former radicals such as Wordsworth and Coleridge, were uncomfortable with this new world vision.

In the days of their forefathers revolutionaries had always looked backwards, to some mythical golden age which at least offered the comforts of familiarity and honour.

The Prometheans would begin with a fresh canvas. They would raze the old world to the ground and begin again. No wonder reactionaries feared them. Perhaps they were right to: Shelley's direct descendant as the cult's prophet was Karl Marx.

Shelley's four-act manifesto *Prometheus Unbound* was written two years before his death after the splenetic and deadly political rallying cries 'Sonnet: England in 1819' and *The Masque of Anarchy* had been suppressed by his friend and publisher Leigh Hunt for fear of imprisonment. Untainted by England's political grubbiness, Shelley's Prometheus was 'the type of the highest perfection of moral and intellectual nature impelled by the purest and the truest motives to the best and noblest ends'. 'We will take our plan from the new world of man,' he chorused, 'and our work shall be called the Promethean.'

Shelley failed his fellow Prometheans. He was not in England when he was needed. His political activism was theoretical, not practical. Even when he lived in London he was not part of the Promethean set. He was self-absorbed and self-destructive. And he died too young. The surviving firebringers – and they include J. M. W. Turner, Michael Faraday, Elizabeth Fry, Charles Babbage, the Brunels, Caroline Norton, the Hunts, Henry Brougham and others – must now fulfil Shelley's vision.

The Prometheans had much to do and their greatest achievements lay before them as Shelley's body lay burning on the beach at Viareggio. If the prophet of the cult was now dead, his high priest, at least, lived on. The temple at which many acolytes now gathered was the Marylebone house of the Northumbrian artist and engineer John Martin, who single-handedly invented, mastered and exhausted an entire genre of painting, the Apocalyptic Sublime. A radical and a subversive, Martin enjoyed a stellar career in the face of critics, swindlers and the disdain of the art establishment. The subversive message behind his paintings went almost completely unrecognized.

At the Martins' house in Marylebone many Prometheans gathered socially. Here one might meet the finest singers and poets of the day, play whist with ageing revolutionaries like William Godwin, admire the beautiful and scandalous Caroline Norton and be treated to a demonstration of Professor Wheatstone's new telegraphic machine – or, indeed, to one of his many practical jokes. Here the young Charles Dickens could rub shoulders with Isambard Kingdom Brunel, Michael Faraday or even Shelley's widow. It was a house of laughter and fun, of passionate political debate, of art, science, music and literature.

The boundaries which later generations of artists, writers, scientists and engineers erected to protect their interests barely existed for Martin and his friends. Most had been born before such professions existed. Many played a part in creating them. The bonds that united these people – their interest in everything, their desire for social justice, their sense of fun and drama, their love of mechanical and natural beauty, their hatred of tyranny – were stronger than self-interest. That they envisioned a new, better world and set about building it for their children is beyond question. That such a glittering band of men and women emerged in a single generation needs some explaining.

CHAPTER ONE

The chains of slavery

NEWCASTLE UPON TYNE. Saturday, 23 November 1771.

> Yesterday se'nnight it began to rain here, and continued without
> intermission until Sunday morning, but this was not to be
> compared to what fell at the head of the rivers Tyne and Wear and
> Tees, which caused so great a land flood as there is no traces of,
> either in memory or record. The Tyne began to rise here about
> twelve o'clock on Saturday night, and about five in the morning
> the arches of the bridge were filled with water ...
>
> Soon after the arches were filled, that one North of the Toll
> shop was entirely swept away, and another on the South end also
> fell, with eight houses on the West side of it, those on the East side
> are still hanging by their timbers. And on Monday, about four in
> the afternoon, another of the South arches fell, with the houses on
> it, and the whole is so much shattered that it must be entirely taken
> down. The loss to the inhabitants is very great, and seven persons
> fell with the houses, and were drowned.[1]

The reaction of the townsfolk and burgesses to this natural disaster was
pragmatic. Within a week of the great flood a 'commodious and safe
FERRY for conveying carriages, horses &c across the River Tyne'[2] was
in service, and although two more houses on the west side fell into the
river, advertisements were already being drawn up for a competition

to replace the many bridges destroyed across the northeast counties of England. Within eight months the first pile for a new, temporary wooden bridge at Newcastle was driven into the soft mud of the riverbed and less than a decade after the flood a new stone bridge, designed by John Smeaton, was opened to traffic. This was a triumph of man's determination to overcome the savage power of nature. Even so, at the time of the deluge correspondents of the *Newcastle Courant* were quick to point out what must have seemed obvious to many: 'We have of late been alarmed with very dreadful symptoms of the displeasure of heaven, by terrible inundations of the rivers in these northern parts.'[3]

The *Courant*'s readers might be forgiven for having thought that the whole world was in turmoil. Across Russia, they read, tens of thousands of people were dying of a fever, probably typhus. From Corsica there were reports that the French had suffered five hundred casualties in their recent action with the 'mountaineers' of that island. There was also notice of a dispute with Spain over possession of the Falkland Islands, ceded to Britain by treaty earlier that year. And if anyone doubted the values of British sobriety – 'our sullen resistance to innovation, thanks to the cold sluggishness of our national character' as Edmund Burke[4] put it, there was an even more sinister item to digest: 'A dangerous conspiracy has been found out at the Seraglio at Constantinople, against the Grand Signor himself, which was discovered by a woman. Though nothing passes in the Seraglio that can be fully known without doors, yet it is judged from the number of dead bodies floating without the walls of the Seraglio, there were many conspirators.'[5] There were, however, more comforting and parochial items in the press. A 'Sober PERSON' was required by an attorney to fill the position of clerk. Elsewhere a Mr Rathlan, Fellow of the Royal Society of Göttingen, announced his arrival at Sunderland with intent to practise there as a man-midwife. And there were the usual advertisements:

> WOOD'S Cephalic TOBACCO: It gives immediate ease, to nervous Head-achs, or Head-achs of any kind; chewing or smoaking cures the Scurvy in the Gums; prevents pain in the Teeth; warms and comforts a cold windy stomach ...
>
> One shilling a cannister.[6]

Tobacco aside, these items would have been of special interest to at least two visitors to Newcastle that year. One was General Pasquale Paoli, who went to dine with the Duke of Northumberland at Alnwick Castle in September. Paoli was the Corsican resistance leader whose progressive parliamentary constitution prefigured those of revolutionary France and America. In 1769 Paoli lost control of the island to France at the Battle of Ponte Nuovo and fled to exile in England but he must have been heartened to know that his fellow patriots were making life hard for the French invaders. He may have wondered, at the same time, how his old secretary Carlo Buonaparte and his young family were managing back in Ajaccio.

There was a second, even more significant revolutionary in Newcastle in 1771. Jean-Paul Marat had been living there for some time, practising medicine, conducting scientific experiments on light and the electric fluid and patronizing the libraries, bookshops and coffee houses of the Bigg Market. Marat was a paradoxical character obsessed, according to one biographer, by 'a morbid expectation of unjust treatment'.[7]

While he was in England, Marat formed ideas which he first published in three volumes between 1772 and 1774. A copy of the last of these, the infamous *Chains of Slavery*, still survives in the library of the Literary and Philosophical Society of Newcastle upon Tyne. Marat later claimed Britain's constitution as a model of political wisdom but the introduction to *The Chains of Slavery* gives a taste of the splenetic invective that made him notorious as one of the Jacobin architects of the French Revolution, the self-styled 'Ami du Peuple'. Addressing the British electorate he wrote,

> If by collecting into one point of view under your eyes the villainous
> measures planned by Princes to attain absolute empire, and the
> dismal scenes ever attendant on despotism, I could inspire you with
> horror against tyranny, and revive in your breasts the holy flame of
> liberty which burnt in those of your forefathers, I should esteem
> myself the most happy of men.[8]

Unlike Jean-Jacques Rousseau, who read into the myth of Prometheus the warning that those who played with fire would get their fingers burnt[9] Marat saw fire as a liberating weapon, a tool of insurrection and the cleansing of sin. The British government naturally enough attempted to suppress *The Chains of Slavery*. Marat discreetly left the country and travelled in Holland for a while, before returning to the northeast of England. Here he visited friends belonging to various Patriotic societies, distributed copies of his works, and only returned to France in 1777.

Newcastle had its own radicals but it was overwhelmingly a mercantile town, its wealth coming from exports of coal, glass, salt and ironmongery. Its citizens did not yet lead the world in locomotion, shipbuilding, glass-making, light-bulbs or naval armaments, as they did a century later. Physically the city was little changed from the days when it was a medieval border garrison. Its ancient walls had been manned as recently as 1745 when Charles Edward Stuart, the Young Pretender, sought to take it and hold London's coal supply to ransom.

Apart from the devastation meted out to bridges, the flood of 1771 caused terrible damage further up the River Tyne: at Bywell, where a church had stood since before the Conquest, corpses were washed out of their graves. At Wylam, where there had been a mine as early as the thirteenth century, the entire colliery was flooded. In the late 1700s the collieries of the northeast were about to embark on a revolution every bit as significant as the political upheavals brewing in America and France. Their futures were, in a sense, forged together. In America and France revolution, when it came, was bloody and prolonged. In Britain it passed unnoticed for a while, until the sound of wheezing engines and coughing children became too loud to ignore.

Flooding in coal mines was an old problem, one that got worse as coal seams inevitably came to be won, exploited and exhausted at ever greater depths. It was controlled by the use of steam-driven pumps whose engineers must continually improve their output and efficiency to keep the costs of pumping under control. The two other serious problems facing coal owners in the late eighteenth century, firedamp and transportation, were more intractable. Firedamp was the deadly gas methane, the cause of many an explosion underground and the deaths of hundreds of men and boys.

The story of the battle against firedamp in the next decades is one of gradual progress punctuated by disaster and setback. Less spectacular but no less intriguing is the story of how coal was transported from pit-head to seagoing collier. The traditional method in the northeast coalfield was to hand-load coal piece by piece into wagons pulled along wooden tracks by horses. By the second quarter of the eighteenth century, long before locomotives were invented, scores of these wagonways laced their way across the hills and denes of Durham and Northumberland.

No one thought wagonways were a perfect solution to bridging the ever-increasing distance between pit-head and river, where coal was loaded, again by hand, into flat-bottomed boats called keels[10] and ferried downstream to where seagoing colliers could moor. It took war and a shortage of horses before colliery owners began to look seriously for a new mode of propulsion. That a steam-powered solution should interest them is no surprise, considering the mountains of coal they could lay their hands on.

Steam power had been around for most of the century. By 1770 there were already more than a hundred and fifty of Thomas Newcomen's atmospheric engines at work across Britain. Most of them were being used to pump water: from mines to keep them dry and from streams into mill ponds to keep their wheels turning – which seems rather ironic. The true potential of steam would not be unleashed until the engine was released from its static prison, so that it could either be moved to where it was needed or used to transport heavy freight like coal. That did not happen until a man born in 1771, the Cornishman Richard Trevithick, mounted a high-pressure engine on wheels exactly 30 years later.

Natural water power for mills and small workshops was long-proven technology with a fatal disadvantage: that suitable streams tended to dry up in summer and freeze in winter (by 1771 the 'little ice-age' was very nearly over and the last frost fairs took place in 1814, but at the time no one knew it). Water power was also paradoxically static: it had to be used where and when it was available. It was Richard Arkwright, a peruke-maker cum publican, whose ideas for spinning cotton on a grand scale exploited the full potential of water power and unleashed the factory system on the world. And he built his revolutionary factory at Cromford in Derbyshire in the year of the flood: 1771.

Cromford was not the first great factory mill but it had a galvanizing effect. It proved that raising capital to pay for machinery and buildings was worthwhile for both borrower and lender; that technology was something worth wealthy men investing in; and that the exploitation of cheap labour was more profitable when that labour was concentrated in one place under the control of a single master.

The factory system as Arkwright envisaged it is not an attractive one. To modern minds the long hours, low pay and virtual bondage of the workers make a striking contrast with the idea of the homespun cottage industry of earlier generations. It was an image exploited by political reactionaries from William Cobbett to Karl Marx and Friedrich Engels. The truth is that rural poverty and deprivation existed before, during and after the introduction of the factory system. Nevertheless, from the hand-loom weavers of Manchester in 1791 to Luddites and the Swing rioters of 1830, labour-saving machinery provoked an aggressive response and was a potential source of unrest that governments feared might ultimately lead to revolution.

Those who lived through the end of the eighteenth century and the first half of the nineteenth century witnessed an unprecedented era of change. Looking back, William Makepeace Thackeray wrote wistfully, 'We who lived before railways and survive out of the ancient world are like Father Noah and his family out of the Ark.'[11] Their childhoods belonged to an age which moved at the pace of the horse and the sailing ship. Their laws counted more than two hundred offences as deserving of the death penalty. Man was 'born free, but is everywhere in chains'.[12] Of a population of something like six and a half million people the vast majority still worked on the land as ploughmen, shepherds, threshers, or in related cottage industries. Land, and therefore wealth, was concentrated in the hands of a minority whose incomes might be thousands of times greater than that of their servants and labourers. The electorate was tiny, parliamentary seats distributed so ludicrously that the bustling manufacturing town of Manchester returned no members at all to Westminster while the abandoned ancient hill fort of Old Sarum in Wiltshire returned two. But many children born into this age of extremes lived long enough to see the Great Exhibition of 1851, the publication of the Communist Manifesto and their world moving at a pace limited only by the potential power of the steam engine.

These children, many of them the sons and daughters of farmers, blacksmiths, parsons and squires, were the first to envision a new world and then fashion it. They invented the professions, the modern notion of rationality, the concept of public health. They abolished slavery and codified rights for women and children. They fought for the freedom to speak without fear of persecution (and sometimes they won). They gave us locomotives, iron buildings and unpickable locks. They endowed their children's futures with electricity and chemistry, with photography and cheap printing, the computer, the telegraph and the idea of education for all. They achieved all this in the face of a corrupt, conservative and deeply entrenched establishment. No wonder they venerated Prometheus, the Greek Titan who stole the secret of fire from Zeus and gave it to man, only to be chained to a rock in perpetual slavery as punishment. He was the libertarian's martyr.

For Fenwick Martin, born in Northumberland's Tyne Valley in the year 1750 during the reign of King George II, freedom existed in an immediate, physical sense. He was a restless man, feckless perhaps, unable or unwilling to spend his whole life in one job. It was not that he lacked talent. He was expert in the art of tanning leather and enjoyed a reputation for being one of the finest swordsmen of his day. In his own way he was a good father too, encouraging and supportive of those of his children who survived infancy. But he liked more than anything to roam the hills, tracks and woodlands of his native Tyne valley.

Fenwick Martin's children saw the coal trade, the birth of the railways and a revolution in agriculture transform the economic and political fortunes of the northeast, but for Fenwick himself the future was an unending search for amenable work on the road, in tanneries, taverns or on farms. He was no great respecter of boundaries, physical or social. At one time he was a drover, taking cattle from the great fair at Stagshaw Bank near Corbridge, where in a week a hundred thousand sheep and cattle might pass through on their way south, to the markets in London. It was a highly skilled trade, licensed and regulated, and maybe that was why Fenwick didn't stick at it. Perhaps it was his interest in a local girl called Isabella Thompson that distracted him. Her parents did not regard him as a steady, respectable match for their God-fearing daughter; and nor was he. But Isabella shared his taste for adventure: they eloped together on horseback in the year of the Great Flood and rode to Gretna Green where they were married, in fairytale fashion, at the smithy's anvil.[13] They had thirteen children, of whom five – William, Richard, Jonathan, Anne and John – reached adulthood. It is symptomatic of this era of high infant mortality that the names of the other eight are unrecorded.

The Martin and Thompson genes made an interesting combination. He was a man of huge energy, of 'dauntless resolution', charismatic and physically strong. She, what little we know of her, was fiery, an extreme Methodist prone to visions and fervent in her biblical interpretation of the world. With parents like that anything might happen.

Their first son, William, was born in 1772 while his parents were still living at Bardon Mill, some 30 miles upstream from Newcastle. When he was four years old Isabella's parents, the Thompsons, moved to the tiny fishing village of Killcolmkill at the southernmost tip of the Kintyre peninsular in Scotland. For unknown reasons they took William with them so that his early years were spent in about as remote a part of the kingdom as was possible.

It was not remote enough to avoid entirely the effects of revolution raging across the Atlantic. In 1775, fired by unrest over taxation and an emerging sense of nationhood, the Massachusetts militia forced the British army out of Boston. The raising of the Continental army under George Washington, the *Declaration of Independence* in 1776 and a bitter war of liberation destroyed the transatlantic child–parent relationship. Throughout the American war, which lasted until 1783, Kintyre was awash with rumours that the notorious privateer John Paul Jones, a native of those waters, was poised to attack Campbeltown's fishing fleet. William may have been blissfully unaware. He had his hands full with the local boys, who had him marked out as an outsider until he eventually learned Gaelic and was able to give as good as he got. Later, he learned his father's skill with the sword and as an adult he claimed never to have been beaten. The only portrait of him, drawn in late middle life, shows him to have been florid with slightly puffy eyes, bushy eyebrows and a long, strong nose. He looks as if he had high blood pressure.

William as a child was robust, his grandparents' household comfortable and God-fearing. Their world was peopled by fishermen, lairds, crofters, the Kirk and the odd Redcoat. There was no sense of intellectual enquiry beyond the wistful. Why, after all, should a pig-farmer read Marat's *Chains of Slavery* or Adam Smith's *Wealth of Nations*? This was an empirical world dominated by man's relationship with nature. William's particular passion was for inventions. He had a great instinct for looking at a device and seeing how it might be improved. If he had been born a generation later his parents, or a teacher, might have suggested he study to become an engineer. But in 1780 there was no such profession. Machines were built by blacksmiths, millwrights and wheelwrights and these were trades handed down almost exclusively from father to son.

William never achieved the intellectual or artistic heights of some of his contemporaries; his doggerel was appalling, his pamphlets vituperative and ill-conceived; his pictures clumsy, sometimes bawdy. It was not for lack of role models, because purely by chance he came early to an acquaintance with Robert Burns. William's parents, Fenwick and Isabella, moved in the late 1770s to Ayrshire where Fenwick found a job as tannery foreman at the Bridgehouse in Brig of Doon. William was able to visit them, making the 60-mile journey across the Firth of Clyde by boat, and there he came to know the Scottish Bard, who was a favourite of the family. He later claimed never to have seen Burns sober.[14] Perhaps a drunken poet, however brilliant, was not such a good example to follow.

From the end of the American war until more than a decade later William is lost to sight before turning up working at a ropery at Howdon Dock near Wallsend. By this time he had brothers: Richard, born in 1779 in Ayrshire and Jonathan, born in 1782 at Highside, near Hexham, where Fenwick had found employment as a woodsman and was also teaching swordsmanship. Richard's early days are invisible; we can only infer that he was tall and strong for he later joined a Guards regiment. Jonathan's we know much more about because he wrote an autobiography. His self-portraits from later in life liken him to his father and show a lopsided, thoughtful face with a mouth compressed in disapproval, a strong nose and receding hairline; very much more like William than their youngest brother, the confusingly named John.

Barely literate and betraying the signs of instability that would destroy him, Jonathan's autobiography is nevertheless remarkable. His childhood was full of portents. He was tongue-tied at birth and carried a speech impediment through his infancy. From an early age he seems to have been prone to night-time wanderings among the lead mines of the Tyne valley. More than any of his brothers he was weaned on his mother's extreme Methodism. 'There is a God to serve and a hell to shun, and all liars and swearers are burned in hell with the devil and his angels,' she is supposed to have told him.[15] He inherited her propensity for visions and his seem to have been of the fiery sort. In later life he remembered having been affected by the story of two men caught in a thunderstorm. One of them, trusting to God, had continued with his journey and arrived safely. The other had taken shelter beneath a tree and been struck by lightning. For Jonathan Martin sin and faith were tangible, physically manifested in punishment and reward, like lightning the tools of a wrathful God.

Contemporaries like Humphry Davy and George Stephenson were fascinated by phenomena like lightning because they saw in them sources of nature's limitless power. Whether this power was a manifestation of the Creator was a moot point; what excited these men, the future heroes of science and engineering, was its potential if only it could be harnessed. Other men looked even deeper. For that first and greatest American Promethean Benjamin Franklin there was an unavoidable connection between harnessing lightning and revolution – Marat's flame of liberty tamed, perhaps.[16] The electric fluid was a highly fashionable subject of interest among men and women with enquiring minds.

In the Bible, as Jonathan Martin must have known only too well, the original sin was to eat of the apple of the tree of knowledge. But in the late eighteenth century a powerful, parallel myth offered enlightenment thinkers a more ambivalent and exciting vision of knowledge. This was the story of Prometheus, later hailed by Karl Marx as the foremost saint and martyr in the philosopher's calendar.[17]

Prometheus was the Titan who stole the secret of fire from Zeus and gave it to humans so that they might be superior to the animals and free to choose their own destiny. Zeus punished humans by introducing Pandora, the first woman, along with her accompanying evils: disease, envy and spite. He punished Prometheus by having him chained to a rock in the Caucasus Mountains. Here an eagle tore out his liver, which regrew every day so that he was sentenced to everlasting torment for his temerity until freed by Heracles. Prometheus' fire, which he took from the sun, was not just a symbol of warmth and comfort. Tamed and directed it represented the arts, particularly the arcane mysteries of alchemy and metalworking. Its flame stood for freedom and immortality, its light for truth and progress, its heat for power that unleashes nature's secrets. Joseph Wright of Derby's *Alchymist*, painted in the year 1771, is no shabby sorcerer: he is positively messianic.

Prometheus is an equivocal gift to the poet: he is both the good thief and the insolent iconoclast; a friend, even a martyr to the downtrodden and a threat to the establishment. His flame is not just a comfort against the cold of night and a light in the darkness of ignorance and tyranny; it is also the cleansing, razing tool of vengeance and revolution. Those who play with fire must be careful not to burn themselves or lose control of the element they have unleashed. It is easy to see how politically, artistically and intellectually charged this myth was, not just for the ancient Greeks but for all humanists from the Renaissance onwards. It binds free will, responsibility, progress and guilt in an exquisite, Gordian knot. Prometheus is perhaps to be identified with the Satan of *Paradise Lost*; if so, he was already a potent icon a hundred years before the birth of the Industrial Revolution.

Prometheus was much painted in the seventeenth century (the century of Milton and the English civil war) by Rubens, Jan Cossiers and Salvatore Rosa among others. And in the late eighteenth century, with its forges and furnaces, steam engines, volcanic eruptions (Vesuvius exploded cataclysmically in 1779 and was active for two generations afterwards) and incendiary revolutions, Prometheus was as powerful and ambivalent an image as Karl Marx would be in the twentieth century. Gothic visionaries like William Blake and Henry Fuseli portrayed his suffering as Christ-like sacrifice. Jean-Jacques Rousseau had warned in 1750 that those who played with Prometheus's gift would burn their fingers. For Goethe, who like Shelley saw in Prometheus a purer counterpart to Faust's Mephistopheles and Milton's Satan, he was an atheist's martyr as he was for Marx.

In Jonathan Martin's psychosis every incident was invested with meaning, his view of Promethean knowledge (like Rousseau's and later Mary Shelley's) decidedly negative. He gives one an uncomfortable feeling, like watching someone walking very close to the edge of a cliff. His interpretation of an accident that caused the death of a younger sister was that she must have sinned in some way.

It was not a healthy mental start to a life that lacked any formal education, particularly with such a feckless father and a rabidly evangelical mother. Jonathan claimed he was so traumatized by his sister's death that he was sent away from the family home to stay with an uncle who farmed land near Hadrian's Wall. He spent some years there tending sheep and later wrote, 'this suited my mind very well, as I could retreat into solitude, and meditate on the goodness of God'.[18]

William's and Jonathan's childhoods are not untypical of the first fruits of the Promethean generation. Their education was empirical, their environment undisciplined, their parents (at least one of their parents) liberal, not to say indulgent. But they were growing up in a world that was changing fast. Whether or not they would seize the opportunities that lay before them remained to be seen.

CHAPTER TWO

The rights of man

BY THE TIME THE PSYCHOLOGICALLY crippling war against
America finished in 1783 it had lasted eight years. Its long-term effect
on the British economy proved to be less disastrous than expected;
certainly less than on that of France, whose ministers had mortgaged
the country's economy by their support for the rebels. Instead of taxing
the colonies, George III's sanguinary mercantile subjects proceeded
to trade with them.

The American Revolution may have helped to mask some of the
subtle changes which took place at home in those eight years. In 1776,
the year of the Declaration of Independence at Philadelphia, Matthew
Boulton and James Watt built their first commercial steam engine.
Thomas Newcomen's engine had injected cold water into the cylinder
to condense the steam and create the vacuum necessary for atmospheric pressure to push the piston down. Watt, a scientific instrument-
maker from Glasgow, realized that its inefficiency was a result of the
cylinder being alternately heated and cooled: heated by steam and
cooled by the injection of cold water. Energy was being wasted in the
process.

Watt designed an engine with a separate condenser; he also used
steam to push the piston down from above while a vacuum was created
beneath by condensing the steam below the piston. It was therefore
capable of greater power and used less water and coal than a Newcomen

engine but it required a higher degree of engineering accuracy than any previous design. Watt was granted a patent for it until 1800 and went into partnership with the Birmingham industrialist Matthew Boulton, because the engine's construction was now beyond the capabilities of a smithy workshop. For the first time, steam engines were the product of factory processes. But while the Boulton and Watt design was a significant technical advance, the patent had the effect of stifling any serious competition for a generation. It was not until the patent ran out that Richard Trevithick gave the world high-pressure steam, together with its power of locomotion.

Alongside the steam engine, 1776 was also the year in which for the first time steam was used instead of water-powered bellows to blow hot air into an iron furnace. The man who pioneered this technique was John 'iron-mad' Wilkinson, who two years previously had designed a new cannon-boring lathe to improve the accuracy and strength of naval guns. This lathe was now used to bore more precise cylinders for the new Boulton and Watt engine and it marks a small but important breakthrough in the process of increasing engine power that led to the creation of the locomotive.

Many more minor revolutions happened in parallel with the political upheaval in America. In 1779 Abraham Darby built the first iron bridge at Coalbrookdale and a velocipede was seen on the streets of Paris. In 1780 British coal production hit ten million tons for the first time and the County of Yorkshire petitioned the government to reform its finances and put an end to corrupt political patronage. The following year James Watt patented his rotary steam engine, paving the way for the next phase of the technological revolution. For the first time a new source of power was available: to drive machinery that had previously been propelled by wind and water. Rotary engines could be built wherever coal could be supplied, relieving the dependence of mills like Cromford on water power alone. The steam engine had graduated from a simple pump to become the workhorse of the Industrial Revolution.

Now, between 1783 and 1784, Henry Cort revolutionized the material that would be used to construct the new world. Cort overcame the two most intractable problems facing the ironmaster. One was that by using coke as a fuel, though it was by now cheaper and more widely available than charcoal, impurities were introduced into pig-iron during the smelting process. These impurities made the metal brittle and difficult to work. The other problem was that iron bars suitable for construction had to be hammered into shape or cut into strips while the metal was still hot. So at the eighty or so blast furnaces now operating across the country, production of iron was slow. Cort's process, called puddling, utilized a second furnace that converted pig-iron into tough and malleable wrought-iron by drawing very hot air (using John Wilkinson's steam bellows) across the metal while it was stirred, or puddled. In this way there was no physical contact between sulphurous fuel and molten iron and the hot air gradually drew off all the impurities inherent in the metal. Cort then solved the difficulty of making bars by using a technique borrowed from Swedish steelmakers. He introduced a grooved roller which could turn 15 tons of iron into bars in 12 hours.[1] From the mid-1780s British iron production doubled in a little over a decade, although Cort himself never made any money from his patents. The consequences of this innovation were not lost on interested contemporaries. Lord Sheffield wrote of Cort's ironworks in the royal dockyard at Portsmouth: 'It is not asserting too much to say that the result will be more advantageous to Great Britain than the possession of the thirteen colonies [of America]; for it will give the complete command of the iron trade to this country, with its vast advantages to navigation.'[2]

Intellectual advances ran parallel to those of the early industrial pioneers. The year of the Declaration of Independence, 1776, also saw the publication of Scottish economist Adam Smith's *Inquiry into the Nature and Causes of the Wealth of Nations*. It was a brilliant and rational analysis of how capital and the division of labour combined to create wealth, and it gave impetus to the new manufactory system being pioneered by Richard Arkwright, Matthew Boulton and others. Between its first publication and the beginning of the French Revolution in 1789 the work went through five editions. But although Smith is today revered as the father of free market economics, his intended message was altogether more equivocal. His argument for the distribution of the wealth is something we can recognize as socially progressive:

> Servants, labourers and workmen of different kinds, make up the far greater part of every great political society. But what improves the circumstances of the greater part can never be regarded as an inconveniency to the whole. No society can surely be flourishing and happy, of which the far greater part of the members are poor and *miserable*. It is but equity, besides, that they who feed, cloath and lodge the whole body of the people, should have such a share of the produce of their own labour as to be themselves tolerably well fed, cloathed and lodged.[3]

A less well-known but equally radical work appeared nearly ten years later in 1785: James Hutton's *Theory of the Earth*. Hutton, like Adam Smith and James Watt, was a product of the fertile Scottish Enlightenment. His studies in geology overturned the long-prevailing view of Archbishop Ussher that the earth was created in a seven-day period during the year 4004 BC. Hutton proposed the idea that the earth was an indefinitely old, complex sphere of rock with a molten core. The proof of its great antiquity was that the processes which must have formed the earth's geology were the same as those which could be observed in the present: volcano, earthquake, flood and erosion; that these processes took an immense amount of time to effect large-scale changes; and that they must therefore have been operating for millions, rather than thousands of years. This simple-sounding principle, Uniformitarianism, underpins all modern geological theory. It led to Charles Lyell's work on geology and the antiquity of man in the next century and opened the way for Charles Darwin's theories on evolution. At a deeper level it held implications for science: when a process could be shown by experiment to produce a repeatable effect the result could be used as an analogue for future work to build upon.

Like Hutton's rivers, volcanoes and floods, the processes of scientific reasoning were invisibly eroding faith in existing concepts of knowledge. They were also releasing humans from some of their self-imposed chains of slavery, reinforcing the liberal interpretation of the Promethean myth, that individual acts of self-improvement and sacrifice might bring benefits to all.

The obvious connection between this scientific upheaval and political revolution was not lost on contemporaries. Thomas Jefferson, co-architect of American independence and third president of the United States, made the link explicit: 'All eyes are opened, or opening, to the rights of man. The great spread of the light of science has already laid open to every view the palpable truth, that the mass of mankind has not been born with saddles on their backs, nor a favoured few booted and spurred ready to ride them.'[4] Too subtle for some, perhaps. Jefferson pondered his Declaration fully aware that in *Common Sense*, the first work to openly demand independence for America from Britain, Thomas Paine had this to say about a monarchy that claimed the precedent of divine right from William the Conqueror: 'A French bastard landing with an armed banditti, and establishing himself King of England against the consent of the natives, is in plain terms a very paltry rascally original. It certainly hath no divinity in it.'[5]

Paine, son of a lapsed Norfolk Quaker, published his first political work in England in 1772. It was hardly revolutionary: an article making the case for better pay for excisemen. Like Jean-Paul Marat, he was a gifted amateur scientist and engineer, apparently holding a patent for a single-span iron bridge;[6] like Marat he was later elected to the National Convention of the French Revolutionary government (though he did not speak French). Like Marat his invective cost him his liberty and very nearly, but for a stroke of good fortune, his head.

Paine was not the only English exile to be elected to the National Convention. Joseph Priestley, the isolator of oxygen in 1774, member of the Lunar Society and radical reformer, was hounded out of England by a mob in 1791 even as effigies of Tom Paine were being burned in towns and villages across the country. A founder of the Unitarian movement,[7] Priestley suffered like the Parliamentarians of the seventeenth century from the delusion that Anglo-Saxon England had been governed under some sort of consensual parliament elected by universal suffrage; he was a humanitarian, not a political philosopher.

These English revolutionaries, if that is not too strong a word, were united in a circle of acquaintances under the patronage of the publisher Joseph Johnson. Johnson ought to have been the publisher of Tom Paine's *The Rights of Man* in 1791; but the visit of a number of agents from His Majesty's Government induced him to hesitate while Paine found another publisher and made a sharp exit for Paris. Johnson was, however, the publisher of William Godwin and Mary Wollstonecraft in the 1780s when William Blake was working for him as an engraver. He also published works by Priestley and by Priestley's Lunar friend Erasmus Darwin. He was one of the first manifestations of the role that publishers would play in bringing together radicals and promoting their ideas in the following decades.

The American Revolution had exercised the British Crown, mercantile interests, and philosophers, but there was in reality very little chance of it spreading among the disenfranchised labourers and artisans of England. It was a war of secession, not of usurpation. The French Revolution, of an altogether different hue, exercised the whole nation. It was politically and geographically too close for comfort; sparked by hatred, fuelled by a fragile ideology. Its leaders burned with ambition. In its heady first days it attracted adherents among the young, the idealistic and the romantic, as such movements will:

> Bliss was it in that dawn to be alive,
> But to be young was very heaven! Oh times!
> In which the meagre, stale forbidding ways
> Of custom, law, and statute, took at once
> The attraction of a country in romance.[8]

So wrote William Wordsworth, who went to France in the immediate aftermath of the Revolution as a giddy 19-year-old. In later years he, like his friend Coleridge and to a lesser extent William Godwin, distanced himself from his youthful opinions both because of ideological distaste for the Revolution's outcome and because as a member of the establishment it was sensible for him to do so. Who can blame him?

In any case, the majority of reformist writing in France (Rousseau excluded) in the years leading up to the fall of the Bastille in July 1789 shows that the leading thinkers contemplated, if they contemplated at all, a constitutional monarchy rather along British lines. 'There is yet virtue on the throne,' wrote Jean-Paul Marat in the early summer of that year.[9] Marat very probably only turned to the Revolution as a means of self-expression because of a failure to realize his own scientific ambitions. Several times he had attempted to gain entry to the Académie des Sciences and several times he had been rejected. Even his translation into French in 1787 of Isaac Newton's *Opticks* did not win the approbation of his peers that he craved.

Marat may still have been a political moderate, but he was outlawed in October 1789 for having incited an infamous march on Versailles through his newspaper, the *Ami du peuple*. In a dramatic series of escapades made more dramatic by his own pen he fled to England, where he remained until April 1790. Exaggeration was a fatal flaw in Marat. Although he later denied furiously that he favoured a dictatorship, or that he had ever incited massacres on the streets of Paris, it was hard to argue with his often printed words: 'Five or six hundred heads cut off would have assured you peace, liberty and happiness. A false humanity has restrained your hand and delayed your blows. It is going to cost the lives of millions of your brothers.'[10]

In September 1791 France did become, for a short while, a constitutional monarchy and again Marat, convinced of the National Assembly's corruption, returned to England, where he stayed until the early months of 1792. By this time Thomas Paine had published his *Rights of Man* and was himself in exile, in Paris. Many such works were written in response to Edmund Burke's conservative *Reflections on the French Revolution* (1790). *Rights of Man* was one. *A Vindication of the Rights of Woman* was another, whose author, Mary Wollstonecraft, also turned up in Paris towards the end of 1792; in her case it was not merely thrill-seeking, but a wish to escape her infatuation with artist Henry Fuseli.

Fuseli was one of a number of artists busy redefining the Gothic. Like William Blake, he was a visionary. A Swiss exile, he had settled permanently in England after 1779, marrying and eventually becoming Professor of Painting at the Royal Academy. He had painted the subject of Prometheus even before Robert Potter's first translation of Aeschylus' *Prometheus Bound* into English in 1777. Other contemporaries who portrayed Prometheus in this period include Richard Cosway, George Romney and the sculptor John Flaxman. William Blake's imaginary character Orc is also generally regarded as a portrayal of Prometheus and there is some evidence that Blake identified himself with the thief of fire.[11]

In September 1792 the first French Republic was established. Massacres followed in which Marat's role has often been scrutinized by historians. He now emerged, alongside Robespierre and Danton, as one of the foremost leaders of the Jacobins. In the same month there were elections for the new National Convention. Marat was elected as one of Paris's 24 deputies. In the ballot he came first. Joseph Priestley came second. Tom Paine was elected for the district of Pas de Calais.

By January 1793, after Louis XVI had tried to flee France with his family, no trace of sympathy towards the monarchy was left. Marat wrote: 'Gentlemen, you have decreed the Republic, but the Republic is only a house of cards until the head of the tyrant falls under the axe of the law.'[12] On 21 January the king was guillotined. Among the onlooking crowd was Mary Wollstonecraft, whose own Promethean warning of the previous year must now have seemed disturbingly prophetic: 'Man has been held out as independent of the power who made him, or as a lawless planet darting from its orbit to steal the celestial fire of reason; and the vengeance of Heaven lurking in the subtile flame, like Pandora's pent-up mischiefs, sufficiently punished his temerity, by introducing evil into the world.'[13]

By the spring of 1793 Marat had provoked the Girondin faction of the Convention into open conflict believing, perhaps correctly, that they represented the forces of the counter-Revolution. In April he went into hiding, was captured, tried and acquitted by the Revolutionary Tribunal. Two months later the arrest of the leading Girondins assured the future of the Jacobin faction and was Marat's personal triumph. Now seriously ill, and convinced that he had done his duty by the Revolution, he retired from the Convention to his bathtub, the only place where he could find any relief in his scabrous, malnourished state. It was there, on 13 July, that he was stabbed to death by the Girondine zealot Charlotte Corday. On the 17th she was guillotined for his murder, her fatal intervention immortally execrated in paint by Jacques-Louis David.[14]

Tom Paine was luckier. Having argued against the execution of the king on both political and moral grounds he came under the suspicious scrutiny of the ill-named Committee for Public Safety, was arrested and imprisoned. It is said that he escaped execution by sheer fluke. He was being treated by a doctor and his cell door was open when a guard came along and chalked a mark on it indicating that he was to be guillotined. When the doctor left and closed the door behind him the mark remained on the inside and Paine, smartly rubbing it out, was spared.[15]

The horror with which Louis XVI's execution was received in England was only half-disingenuous. Politically Britain was more concerned with the revolutionaries' external wars and their opening of the River Scheldt to trade with Antwerp than with the overthrow of her old enemies the Bourbons. But the British establishment was increasingly concerned that Revolution might spread across the Channel. The government reacted by suspending *habeas corpus* so that political suspects might be detained without trial. A Police Bill was passed in 1792 providing for the first professional law enforcement agency in Britain, whose officers' remit was primarily to prevent revolution. This was followed in 1793, the year of the Parisian Terrors, by an Alien Act requiring all foreigners to register on entry to the kingdom. By 1795 the radical London Corresponding Society, demanding universal suffrage (among men, not women) and annual parliaments[16] was able to bring a hundred thousand people out onto the streets to urge parliamentary reform. King George III's carriage was stoned.

The Prime Minister, William Pitt, was not slow to implement further counter-revolutionary measures. He had already opened lines of contact and support to French royalists and began to assemble his first international coalition in readiness for the war which began in 1793. A Seditious Meetings Act and a Treasonable Practices Act were passed. Leading members of the Corresponding Society, founded by a shoemaker called Thomas Hardy, were arrested and tried; most were acquitted. Whether this was with judicious political connivance or not is hard to say but their momentum was dissipated and, with war increasingly concentrating the country's minds, Britain's nascent reform movement was shelved for another generation.

Their acquittal still surprises. It is an indication that despite the aristocratic and conservative bent of Britain's judiciary, magistrates were fiercely resentful of political interference. What is more surprising is that the most profoundly threatening work of the time was not suppressed at all. William Godwin's rational and deadly *Enquiry concerning Political Justice*, published in 1793, is now regarded as a founding thesis of intellectual anarchism. Written with deceptive simplicity, it influenced not only his own generation but a Europe-wide movement that spanned the entire nineteenth century. Its thesis, in tune with Hegel's influential dialectic idealism in Germany, was that mankind was perfectible, that given time and proper education a society would develop in which government imposed from above was unnecessary. This was desirable because governments inevitably became corrupt. Godwin's arguments, if naïve, were perceptive and threatening. No government likes to be called corrupt even if it is a truth universally acknowledged. And the idea of educating the masses was the sort of cant that had led to the French Revolution. The notion of perfectibility through progress was not to be tolerated.

With wonderful historical irony William Pitt decided that *An Enquiry* was not worth suppressing because, at three guineas a copy, few men would be able to afford it. He was half right. Instead, gentlemen, shopkeepers and workers alike formed societies to buy copies, and shared them.[17] The book, published by Joseph Johnson from his premises at St Paul's Churchyard in the City of London, went to three editions within five years.

Godwin is not just important as the author of *An Enquiry* and a number of interesting novels; not just as the husband of May Wollstonecraft (1759–97), the passionate but undisciplined author of *A Vindication of the Rights of Woman*; nor even because they were the parents of Mary Shelley, who was to write that defining Promethean tale

Frankenstein. He also, vitally, acted as mentor and inspiration to the radical poet Percy Bysshe Shelley. By the time of their first meeting Godwin had become such an obscurity that zealous young radicals like Shelley, who eloped with his daughter Mary, thought he must have died decades before. But in the early 1790s Godwin was still intellectually, if not physically, an electric presence in London.

He was born in 1756, the son of a dissenting minister of Wisbech. Although he practised as a minister himself for a time, he underwent a profound crisis of faith and became one of the first men of his age to publicly profess himself an atheist. His naturally analytical mind took him on a painful journey which began with self-examination and ended by challenging some of the most cherished conventions of eighteenth-century thought. 'Why should I,' he wrote, looking back on his early life, 'because I was born in a certain degree of latitude, in a certain century, in a country where certain institutions prevail, and of parents possessing a certain faith, take it for granted that all this is right?' [18] A statement like that can only emanate from a mind of rare breadth. It is the voice of the true sceptic – though not of the revolutionary that many of his contemporaries thought (or wished) him to be.

It is not really surprising that during the 1780s when he was living in London, the school which he attempted to set up on his own educational principles was unsuccessful. Few parents wish to pay for their children to be instructed in uncertainties. Perpetually struggling without a regular income, Godwin began to contribute articles to Richard Brinsley Sheridan's Whiggish *Political Herald* and the anti-establishment *New Annual Register*.

In 1791 Godwin for the first time met Mary Wollstonecraft (they apparently took a more or less instant dislike to each other) and Tom Paine. They had been invited to a party given by Joseph Johnson, whose support for them and their fellow radicals was to see him imprisoned and fined. In the same year Godwin began writing his *Enquiry concerning Political Justice* on the back of an enviable advance of a thousand guineas from Johnson which eased his money troubles for a while.

Young visionaries in England were fired by the ideals of the French

Revolution, quick to see its moral imperatives and slow to acknowledge its horrors just as British socialists, ignorant of its unpleasant realities, were eager to see the spread of Bolshevism in the 1920s. Along with Paine, Priestley, Blake and Wordsworth the Revolution worked its magic on Mary Wollstonecraft. Her *Vindication of the Rights of Woman* was published by Johnson in 1792, lighting a torch subsequently borne by Mary Somerville, Jane Marcet, Caroline Norton and, much later, the women who finally won for themselves the right to vote. In Wollstonecraft's day, few men had that privilege.

Apart from quite exceptional women like Caroline Herschel, who was an astronomer, and caricaturist James Gillray's publisher Hannah Humphrey, women who sought professional careers were generally restricted (and some of them still chose to remain anonymous) to teaching and writing. As it turned out, the woman's accepted role as gentle educator to the children of gentlemen, combined with the early nineteenth-century flowering of mass-market book distribution, made women writers, especially of popular science, the most influential communicators of their age. There is something satisfyingly Promethean in that insolent thieving of children's minds.

William Godwin's *An Enquiry* was by no means a Promethean tract. It did not propose or excuse revolution. More profoundly, it exposed the flaws inherent in the political system and laid responsibility for that system on all the individuals which made up society. Far from raising the banner of republicanism based on the power of the worker, Godwin sought to diminish the individual. He was, in a way, anti-Promethean: he did not believe in free will. But his *Enquiry* won him a great many admirers among intellectuals and political activists and frightened an establishment already feeling hot under the collar.

Among the zealots who became Godwin's 'followers' was his contemporary, the illustrator and poet William Blake. Blake was working as an engraver for Joseph Johnson when he wrote an account of *The French Revolution* which developed his own ideas of revolt against authority. His greatest poetry (*Songs of Innocence* in 1789, *Songs of Experience* in 1793 and *Jerusalem*, completed in 1820) is rebellious, prophetic, mystical and deeply Promethean. His illustrations for

Paradise Lost (1808) are among the finest to grace any book. He is unique in having been his own illustrator, printer, publisher and bookseller. After his move to the seclusion of a tiny cottage in the Sussex seaside village of Felpham he even managed to get himself arrested for treason (he also was acquitted). He is credited with having advised Tom Paine to quit the country just in time to avoid William Pitt's wrath. Now recognized as a brilliantly visionary poet and artist, Blake was virtually unknown in his lifetime outside Johnson's elite circle. Many, including Wordsworth, thought him mad.

By 1796, when Godwin and Mary Wollstonecraft met for the second time, the realities of war, the threat of invasion and the dissipation of the English revolutionary movement after the Treason Trials had profoundly altered the political climate in England. Talk of parliamentary or social reform became unfashionable. Godwin and Wollstonecraft were by now admirers of each other's works but Godwin had upset the radicals in London and was now yesterday's news. He became reclusive. During the following year this oddest of couples set up a sort of ménage in Pentonville. Despite their common antipathy towards wedlock, they were married in 1797. Wollstonecraft bore them a daughter, also called Mary. She died 12 days after giving birth, leaving the middle-aged Godwin bereft and alone in charge of a tiny baby. In an age when enlightenment and intellectual liberty clashed perpetually with injustice, poverty and human frailty, even the most profoundly rational man of his time might have been forgiven for wondering which Olympian he had offended.

CHAPTER THREE

Children of the Revolution

THE LAST SURVIVING MARTIN SON, John, was born in 1789, when the family was living at Haydon Bridge in the Tyne valley. He was 17 years younger than William and seven years the junior of Jonathan. John was small and neat, with perfectly regular features and bright, intelligent eyes. In adulthood many men and women found him beautiful, Byronic even, with a charm to match. Nearly a generation younger than his oldest brother, he grew up in a world in which age-old securities and inhibitions were being broken down by industrial and political revolution. His extraordinary career as a subversive yet commercially successful artist might not have been possible in another generation. He twice fought and overcame the laws of indenture to free himself from restrictive master–servant relationships. Insulted by the Royal Academy's treatment of commercial artists he snubbed them and eventually wore their contempt as a badge of honour. He rendered their disapprobation irrelevant.

John Martin forged a series of professional and social connections with the great men and women of his day that had little to do with his, or their, rank at birth. In doing so he came to figure as a central character in what might be called the modern cult of Prometheus, with consequences that neither he nor anyone else foresaw. He survived financial ruin, disgrace and critical contempt and lived to influence the shape of modern London, the Pre-Raphaelite Brotherhood, the novels of the Brontës, Jules Verne and Victor Hugo, and a cinematic vision of narrative that has never gone out of style.

Born into a world in which uncertainty and opportunity went hand in hand, John seized his chance; but he did more than that. He conferred the freedom that he and his friends won for themselves on his and their children. They did not just succeed in stealing the secret of fire from their betters: unlike Prometheus, they got away with it.

John Martin was a child of the Revolution: literally, for he was born in the week that the Bastille was stormed in July 1789. Like his close contemporaries Michael Faraday, Lord Byron and Charles Babbage, he found inspiration in the Promethean myth. For them the theft of the secret of fire was an audacious act of liberation, to be admired and emulated. It symbolized man's desire and ability to release the power of nature, not simply for self-aggrandizement (Napoleon, the arch-Promethean of his age, was perhaps a poor model) but to fulfil the potential of all mankind. It was a vision almost unique to John Martin's generation. But it was not confined to Britain.

The American dream, fostered by a hundred and fifty years of pioneering spirit and moral self-improvement, first envisioned by men like Benjamin Franklin and Thomas Jefferson as an act of moral and fiscal liberation, was Promethean. The Statue of Liberty, given to the United States by France in recognition of a friendship born during the revolutionary era, is a thoroughly Promethean figure, flaming torch and all. But Lady Liberty was not France's first gift to the Americans. During the War of Independence she had supplied arms to the revolting colonies and broken Britain's trade embargo with them. And in 1793 she unwittingly gave New York its first great engineer, Marc Brunel.

In later years the Brunels knew the Martins and their circle well. John Martin bought the house in Chelsea where they lived for many years and rode on the footplate of one of Brunel Junior's locomotives. At the time of John Martin's birth and the storming of the Bastille, Marc Isambard Brunel, born in the same year (1769) as Napoleon Bonaparte and the future Duke of Wellington, was a 20-year-old French naval officer serving in the West Indies. His family had for centuries farmed land around the village of Hacqueville on the plains of Normandy. Marc's mother having died when he was seven, his affectionless father determined a career in the Church for the boy. The boy was not interested. Things like cartwheels interested him; cartwheels and numbers.

Brunel later remembered having seen in Rouen, from the banks of the River Seine, the unloading of two giant cast-iron cylinders from a ship.[1] These, he was told, belonged to a steam pumping-engine being imported from England to power the Paris waterworks. Nothing like this was being made in France at the time and the sight of such overwhelmingly massive castings made a great impact on Brunel. He quietly nurtured a desire to see England and went through the priestly motions. He was rescued from the seminary at Rouen by his cousin, a M. Carpentier, a well-connected former naval captain who fostered him, taught him Euclid and the rudiments of trigonometry. In 1786 Carpentier used his political influence to procure for Brunel a berth as a *volontaire d'honneur*, or midshipman, on a frigate bound for the West Indies station. That was how he missed the excitement of 1789. But he cannot have been unaffected by the Revolution: the navy's aristocratic officer cadre was being purged by France's new government, and political commissars would soon be looking over admirals' shoulders.

Brunel returned to France and was paid off in 1792. He soon witnessed the instability of the political situation for himself. In August in Rouen he and others went to the aid of a group of the king's Swiss mercenaries fleeing from the massacre at the Tuileries and were nearly lynched by the town's mob.

While Marat's fellow Parisians led the Jacobin cause, royalist enclaves in the provinces were plotting a counter-revolution that eventually broke out in the south of the country the following year. Normandy, and Brunel's connections, were staunchly monarchist. That winter Carpentier took Marc to Paris to see for himself what was going on there. Brunel's relationship with the Revolution was brief but dramatic. In January 1793, during the days when Louis XVI's fate was being decided in the Convention, Brunel found himself in a vituperative argument with a group of Jacobins at the Café de l'Echelle under the colonnade of the Palais Royal. He could not have provoked the Jacobins more than by calling loudly to his dog, 'Viens, Citoyen!' Then he openly threatened Robespierre with a Norman vengeance to remind him of the Norsemen of old and was lucky to escape when a diversion was created by a sympathetic Girondin deputy. On 17 January he and Carpentier managed to slip out of Paris just as barricades were being set up to isolate the city.

As the Terrors of 1793 spread outwards from Paris, Marc knew he had made himself a target for those parties of sans-culottes who were ranging the provincial towns rounding up Girondin sympathizers. So Carpentier arranged with the American vice-consul at Le Havre to get him a passport and a passage to America. During his dramatic flight Brunel somehow lost the passport after a fall from his horse in which he was injured. When the *Liberty*, bound for New York, was searched by officers from a French frigate, Brunel had to forge himself a fresh document. He landed in New York unscathed but left behind in Rouen his fiancée, a young Englishwoman called Sophia Kingdom whose family had untimely sent her to Normandy to learn French. She was interned in the Gravelines convent in Rouen and not released until after Robespierre had suffered the vengeance not of the Norsemen, but of his own guillotine.

Marc Brunel stayed in America for six years. At first he gravitated towards the expatriate French enclave at Philadelphia, where he met Talleyrand in the back parlour of Moreau de St Mercy's bookshop. In this shiftless state he and a couple of friends decided to embark on a project to map and sell to colonists large tracts of the unexplored territories of the Great Lakes region. During one of these expeditions he had a chance encounter with Louis-Philippe, refugee Duc d'Orléans and future King of France. At another time Brunel and his friends conceived the idea of a canal to link the waterways of the St Lawrence and Hudson with the Great Lakes and they were engaged to conduct a survey for the scheme by a wealthy American merchant. This vastly ambitious project was not completed in Brunel's lifetime; but within a few years he was established as an architect and eventually found himself Chief Engineer of New York City and an American citizen. He improved the defences of Long Island, built a cannon foundry, and competed successfully for the design of the new Congress building at Washington, though it proved in the end too expensive to build. It is hard to imagine that he can have had much leisure time.

Early in 1798 Brunel dined in New York with Alexander Hamilton, a colourful journalist, lawyer and soldier who had been the first Secretary of the US Treasury. His political career was ruined in 1795 after a scandalous affair and in 1804 he was to die from a wound sustained in a duel. But in 1798 he was still a powerful figure, an industrialist of great vision and an intimate of George Washington. At this dinner there was an animated discussion about the recent British naval victories at Cape St Vincent and Camperdown. At Cape St Vincent Sir John Jarvis' inferior squadron had attacked the Spanish fleet and, led by a daring attack from Commodore Nelson, beaten them. At Camperdown, in the immediate aftermath of two serious naval mutinies, Admiral Duncan had brilliantly beaten a Dutch fleet off the coast of Holland by breaking the enemy line in two places and forcing a bloody mêlée. Britain's naval strategy and battle tactics must have made for lively discussion in a country that was just building its first navy.

Brunel, Hamilton and their companions agreed that Britain's naval war machine was an impressive industrial enterprise. Another Frenchman present, who had seen them, described to the company the operations at the English naval dockyards at Portsmouth, and in particular the shop where pulley-blocks were made. Brunel thought the operation sounded rather inefficient and sketched some ideas for improvements.

Within a year he had decided to travel to England and, armed with a letter of introduction from Hamilton to Earl Spencer, the First Lord of the Admiralty, try his hand at block-making.[2] He arrived at Plymouth in March 1799 and went first, not to the Admiralty, but to see Sophia Kingdom's brother. Marc and Sophia were married in November. The block-making contract took longer.

The Revolution and the war between Britain and France were shaping the futures of a generation of men and women. Some, like Brunel, exploited the entrepreneurial potential of the conflict; some were inspired by the thought of glory to enlist in the army; others, men who had been to sea in their youth, perhaps, were pressed into the navy; fewer volunteered. Thousands of women waited at home, paying Jane Austen's 'tax of quick alarm'. Some of these decided to educate themselves and although the professions were denied to them it was beginning to be possible for women to sustain a career through writing.

Mary Wollstonecraft, who died in 1797 shortly after giving birth, left that legacy to her daughter, Mary Shelley. Mary Somerville, whose father was a naval officer, was in her late teens and had already mastered Latin and Euclid. She later translated into English the five-volume *Traité de mécanique céleste* of the French mathematician and physicist Pierre-Simon de Laplace.[3] According to the novelist Maria Edgeworth, Laplace thought Somerville was the only person who really understood his work. She observed admiringly that while Somerville's 'head is among the stars, her feet are firm upon the earth'.[4] Elizabeth Fry, who was born in the same year as Somerville and who had once openly worn a tricolour in the streets of Norwich, was now applying her Quaker credentials to the needy.

Jane Austen, in her early twenties, was quietly, patiently observing the mores of her genteel circle. In 1797 she began to write *Sense and Sensibility* and completed a story which she called *First Impressions* (later recast as *Pride and Prejudice*). The first great modern novels in English, they were finally published in 1811 and 1813 respectively. During her short writing career Austen remained almost unknown, as socially isolated in her Hampshire rectory circle as she was intellectually liberated. All her novels were published anonymously.

One glaring exception to the rule which excluded women from the professions was the astronomer Caroline Herschel, who enjoyed the rare privilege of receiving a salary from the Crown. In 1797 she presented the Royal Society with an index to Flamsteed's famous observations which included hundreds of stars that the Astronomer Royal had missed.

Just as the stars in the heavens, multiplied by the power of telescopes, were revealing themselves as more numerous, so it was noticed that the human population was growing. Britain was supporting more than ten million people by the time of the first census in 1801.[5] But if the planets were governed by Newton's divine clockwork and the economy worked like an engine, what mechanisms governed population? Following Adam Smith's *Wealth of Nations* there came in 1798 a shattering prognosis of unbridled population growth, unsustainable societies, misery and vice. The apologetic author of this awful vision was an English country parson, the Reverend Thomas Malthus. His analysis of an inherently imperfectible society (the antithesis of Godwin's vision) created ripples that were still being felt by Charles Darwin 60 years later. He saw, or he thought he saw, that population tended to rise geometrically, whereas subsistence could only expand arithmetically. He was wrong, but he managed in the midst of war and hunger to scare himself and a great many other people with his anti-Promethean message:

> This natural inequality of the two powers of population and of production in the earth, and that great law of our nature which must constantly keep their effects equal, form the great difficulty that to me appears insurmountable in the way to the perfectibility of society. All other arguments are of slight and subordinate consideration in comparison of this. I see no way by which man can escape from the weight of this law which pervades all animated nature. No fancied equality, no agrarian regulations in their utmost extent, could remove the pressure of it even for a single century. And it appears, therefore, to be decisive against the possible existence of a society, all the members of which should live in ease, happiness, and comparative leisure; and feel no anxiety about providing the means of subsistence for themselves and families.[6]

Little wonder that a primary impetus for technological progress throughout the war was a shortage of food, and not just because of population growth. It was due partly to a series of poor harvests and partly to the demands of the military. Britain's navy in particular was a vast industrial enterprise comprising somewhere between five hundred and a thousand vessels,[7] capable of mobilizing more than a hundred thousand men and keeping them at sea for long periods. Fed on bread and biscuit, on beef, pork, peas and beer, cheese and suet, the Royal Navy's sailors consumed more food than Britain could possibly produce unless she underwent an agricultural revolution.

William Pitt, already in his tenth year as Prime Minister,[8] was not slow to react to the start of the war. In its first year, 1793, he established a Board of Agriculture under Arthur Young, a proselytizer for enclosure whose role was to review agricultural practice across England and Wales and recommend improvements. The king himself, 'Farmer' George III, was an enthusiastic supporter of research into livestock and crop improvements. The Society of Arts awarded medals to men who designed such wonders as turnip drills. On the fertile plains of Norfolk, crop rotation and liming gave rise to improving yields that were both visible and profitable.[9]

Some of the most radical and enduring changes in production took place in the North, among the valleys where the Martin brothers were growing up. Here, in the 1790s, Scottish millwright Andrew Meikle's revolutionary grain-threshing machine was taken up with enthusiasm because coal-mining was creating an agricultural labour shortage. The ideas of the Leicestershire sheep breeder Robert Bakewell had been known for a decade before the war started and they now spread among a group of enlightened northern farmers who saw economic potential in the conflict. Among the most innovative were the Colling brothers of Ketton, near Darlington, who had visited Robert Bakewell in about 1783. From him they learned new techniques for selective line-breeding and brought them home. They proved to be receptive pupils. Their shorthorns became world-famous after 1796 when a series of experiments resulted in the birth of a giant steer known as the Durham Ox. At the age of five years he weighed almost two tons and was bought by a Mr Bulmer for £140. Bulmer paraded him around the country at exhibitions and shows, between which the ox travelled in a specially constructed vehicle.

The Durham Ox was eventually slaughtered, having dislocated his hip, in 1807 when he weighed 220 stones, about a ton and a half.[10] Dozens of public houses were, and still are, named after him. He was more often painted, it is said, than any human contemporary. His fame induced a frenzy of 'fat cattle' breeding (and some wicked caricatures of their equally fat breeders by James Gillray) the progeny of which was consumed by the navy's sailors at a rate of four pounds per week per man.

Elsewhere in the northeast of England other offspring of the Revolution were soon to be affected by the war. In 1794 William Martin was working in a ropery at Howdon, near Wallsend, when the keelmen on the River Tyne set about destroying the staithes built along its banks. These jetties were beginning to incorporate elaborate and ingenious devices to funnel coal from wagons directly into the hulls of coasters without damaging the coal, rather than load them into 'keel-boats' up-river.[11] Keelmen, quite rightly, saw in the staithes (indeed in any technical innovation) their own demise.

This was also a year when 58 men and boys died in underground explosions at local collieries.[12] William Martin was only too familiar with the noise and smoke of pits, the toll of dead hewers and the maze of wagonways that ran along the north bank of the river. It was now that he conceived the first of many inventions. Unlike his youngest brother John, William was not entrepreneurial in the least; his fertile, indeed over-fertile mind simply produced ideas as verbal sketches ...

> When my comrades and I were taking a walk, some of my acquaintances were engine-wrights, and seeing that the coal waggons ran on wood-rails, I would often say to them that I thought the people of the coal trade were foolish for having the waggons run on wood ... I told them I would order cast metal rails to be made and laid upon stone, and make them join each other; then the engines would go with less friction ... and one horse would draw as much as four.[13]

The coal owners were not entirely foolish. Wagonways may have been inefficient, but they were relatively cheap to lay and simple to maintain. The price of fodder for horses was not yet at the level at which it began to cripple them; horses themselves did not become prohibitively expensive until the British army became involved in the Peninsular War after 1808. But William Martin was right to think about the technology of the rails, because it was this technology which held back the progress of the locomotive along those very river banks 20 years later. In the meantime, William's atavistic restlessness urged him to change his own direction:

> In 1795 I went up to Hexham to see my father and mother, as they lived there at the time. My brother John, the historical painter, was a child in frocks, and my brother Richard was serving his time with Mr. Thomas Graham the currier, at Hexham; and at that time I went into the Northumberland regiment of militia under the command of Colonel Reed, and joined them at Durham.[14]

Very soon William made a name for himself in the militia: by beating all-comers as a swordsman and 'leaper'. Having been taught by his father, an acknowledged expert with the sword, William relished taking on challengers, claiming that he was never beaten in a swordfight. Evidently a tremendous athlete, according to his own account he made a standing jump of 12 feet 4 inches which, if true, was somewhat longer than the last identifiable world record in the first decade of the twentieth century.[15] His technique was that of the classical Greek athlete: he held a three-pound stone in each hand and threw them backwards as he leaped.

The militias functioned as a sort of Home Guard, trained and equipped to a relatively high standard. There was a ten-guinea bounty for volunteers agreeing to serve for three years, and it may be that William signed up to pay off some of his father's debts. Discipline was harsh, much more so than in the Royal Navy. By the end of the war it was a cause célèbre among radical journalists. A private absent from morning parade might be given a hundred lashes.[16] In the case of the Northumberland militia, first raised by the Lord Lieutenant of the county in 1759, their deployment was along the east coast of England, between Newcastle and Colchester. Where there were no existing barracks (Newcastle, for example) the troops were billeted on local townsfolk. In summer they protected crops and helped with the harvest. But they were not an entirely benign force. In a country with no formal police service the government regularly stationed them near towns where unrest might be expected. In 1812, at the height of the Peninsular War, there were as many troops stationed in Nottinghamshire to quell Luddite riots as there were in Spain and Portugal.[17]

During the panicky atmosphere that prevailed in the mid-1790s the Mayor of Newcastle, Henry Rudman, wrote to the Home Secretary, Henry Dundas, asking for urgent funds to build a barracks in Newcastle, with this warning: 'The present turbulent disposition of the sailors of this port and the tumultuous spirit which, on several occasions, has recently shown itself among the pitmen and others employed in the coalworks appear to us strongly to enforce the propriety of adopting this measure.'[18] The Mayor got his money, but not until nearly ten years later. In the meantime troops were called out more than once to disperse striking keelmen and rioting sailors. Apart from the imposition of billeting and the threat of suppression, the militia also posed those risks common to troops who are underemployed and have access to drink. They themselves rioted, typically with the men of other militias when they were billeted together.

Within another couple of years Richard Martin joined the army too. He became a fugleman [19] with the Northumberland Fencibles in Ireland during the risings there and remained in the army for 22 years, serving in the Peninsular War and at Waterloo with Wellington, when he was not sponging off his brother John in London. Regrettably, his autobiography has been lost. The only work of his that survives is an extended poem called *The Last Days of the Antediluvian World*, which has a frontispiece by John.

John Martin's childhood was altogether more straightforward than his brothers': he went to school. While Richard and Jonathan served apprenticeships in Hexham, the former to a currier and the latter, after his early traumas, to a tanner, John was sent to Haydon Bridge Grammar school, a free establishment with a progressive reputation. One of its teachers in the 1770s, though he left before John arrived there, had been Thomas Spence, the radical educationalist and reformer born into extreme poverty on Newcastle's Quayside, whose contribution to the revolutionary movement was to propose the nationalization of land. Spence, who probably met Jean-Paul Marat while he was living in Newcastle, was twice imprisoned for sedition for selling Thomas Paine's *The Rights of Man* from a barrow on the streets of London in 1792. His later adherents included the would-be cabinet assassin Arthur Thistlewood. Even today, the northeast boasts its Spenceans.

That John was the first Martin to attend school is not to say that his parents' circumstances had much improved. They were, in fact, living in poverty in a one-room cottage on a farm called East Landends at Haydon Bridge. William had now left, but John, his older sister Ann and his two other surviving brothers were all living in cramped conditions at home. They were at least settled. The school was as much as anything a cheap form of childcare.

Like his confusingly named brother Jonathan, John was prone in his early youth to wandering in the woods and among the ruins of the Tyne Valley. He remembered himself as being timid and nervous, afraid of the dark and of hobgoblins. Unlike Jonathan he digested his mother's Methodism leavened with Greek, Latin, geography, mathematics, writing and navigation. Their stable circumstances gave his childhood solid foundations. He may have shared their visions and their wilfulness, but school rules taught him discipline and education cured him of their fatal naïveté.

In common with his brothers John wrote unaccomplished verse and drew and sketched obsessively. But where William saw a piece of machinery or equipment and instantly visualized how to improve it, John could look at a drawing and see how it might be made better, more lifelike. It looks rather as if his first drawing master was Richard who, John later claimed, was good enough to have become a professional artist himself. Jonathan, whose surviving pictures were executed within the walls of Bedlam, created manic images that are nevertheless both striking and technically clever – reminiscent, some of them, of William Blake. But the brothers all saw that John was born with the most talent and they encouraged him, as did their father Fenwick. Encouragement, or at least paternal indulgence during youth, was a strong feature among the Prometheans, as much a liberating influence as the prevailing spirit of the age.

The Martins were too poor at this time to buy their talented son pencils, brushes or paints. John used every material he could get his hands on, from charcoal to burnt cork, soil, watercolours and school pencils. He used often to wander down to the banks of the Tyne and draw pictures in the silt with a stick. His book of learning was his native landscape: the rivers, hills, oak woods, green pastures and enclosed arable fields of west Northumberland. The ever-changing sky provided his aesthetic palette, just as it did for his future friends, J. M. W. Turner and John Constable.

As an adult John was the only member of the family to indulge in radical secular politics, perhaps initially as a result of his formal education but more probably because he was plunged into reformist circles on his arrival in London. At first glance it is hard to see subversion in his ethereal landscapes and grand historical subjects, but that is because to twenty-first-century eyes his works look 'old-fashioned', the themes irrelevant. To his contemporaries they were shocking. If there had been a law on seditious art John Martin and his friend George Cruikshank might have found themselves imprisoned, as many of their literary friends were. One might compare the public effect of John Martin's pictures to the impact of the early cinema; and it is no coincidence that one of its greatest exponents, D. W. Griffith, used Martin's paintings as set designs for his epic scenes.

Although born in revolutionary times John Martin and his fellow firebringers were not always in the frontline of revolutionary change. By and large they were not, like William Pitt or Horatio Nelson, mercurial (Turner was a quite exceptional prodigy). In many cases they were slow developers. In 1798 the political and industrial revolution was, like Martin himself, still in its infancy. Many of its children were either, as Martin, Michael Faraday and Charles Babbage were, still in their breeches; or, as George Stephenson was, apprentices in their father's workshops.

Richard Trevithick was typical of these paternal apprentices. Like Stephenson he was an empirical spirit, an essentially practical man of little or no formal training or education. As the century drew to a close he graduated from his father's Cornish mine workshop and built his first model of a high-pressure steam engine. He ultimately gave up on locomotives, convinced that public and government alike did not appreciate his engineering skills. In a way he was right: his last locomotives were built several years too early, before a shortage of horses properly concentrated the minds of men of business. Having narrowly failed in his attempt to drive a tunnel under the Thames in 1807 Trevithick concentrated on designing static engines, dissipating his extraordinary talents. Years later he went off to South America adventuring until, penniless, he was rescued by George Stephenson's son Robert. He was perhaps more Icarus than Prometheus.

Why did this generation produce such a seeming glut of single-minded inventiveness and self-promotion? Certainly the wars against France played their part in drawing talent out, as wars always will. But the liberation of the artisan class from their father's shops and smithies is something else. It must in part have been due to a change in population dynamics at the end of the eighteenth century when the birth rate rose and the death rate was static or falling. Before this period of rapid growth sons were expected to follow their fathers into trade or business, to ensure financial protection for mothers, sisters and wives and maintain the continuity of a trade. A father's inventory of tools and clients was not just prized; it was the only means of entry into what was often a closed shop. But when second and third sons began to survive to maturity a family had more options. Male children might, indeed many of them were obliged to, find a means of earning money outside the family business. It was a circumstance that encouraged aspiration and forced young men to look outside the villages where their forefathers had been born.

There is something else. Unlike almost all other European countries at this period, British society was fundamentally casteless. Riven by inequalities in wealth, class and opportunity it was; but entry to the middle class – the class of magistrates, voters, men of business and the clergy – was never exclusively by birth. And a country in which labourer and blacksmith might play hard at cricket with the hereditary squirearchy on the village green fostered the belief among some that a lord was perhaps very much like a peasant when the playing field was flat. In Britain mercantile pragmatism was a force occasionally sufficient to overcome aristocratic snobbery.

Cornishman Humphry Davy, born in Penzance in 1778, was the oldest of five children belonging to a wood-carver. Like Marc Brunel he developed an early taste for a life beyond his father's narrow world. He was precocious at school, and persuaded to undertake an apprenticeship not in his father's unproductive workshop, but with a Penzance barber-surgeon. Here he began to play with the apothecary's collection of chemicals. By 1798 he had already been studying chemistry for a year, reading Lavoisier's *Traité élémentaire de chimie* (the first true chemistry textbook) in French, blissfully unencumbered, through ignorance, by the long-running phlogiston debate.[20] When he was almost 20, his wood-carver father having died, he moved to Bristol to work at Dr Beddoes' famous Pneumatic Institute. Here he met Samuel Taylor Coleridge and Robert Southey, and here he would have to decide whether to follow chemistry or pursue his undeniable talents as a poet; whether, in his own Promethean lines,

> To scan the laws of nature, to explore
> The tranquil reign of mild philosophy;
> Or on Newtonian wings to soar
> Through the bright regions of the starry sky.
>   From these pursuits the sons of genius scan
> The end of their creation, – hence they know
> The fair, sublime, immortal hopes of man,
> From whence alone undying pleasures flow.[21]

Humphry Davy was ambitious. At Beddoes' Institute he learned quickly, met all the right people and was marked out for a successful career. But medicine was not enough. Experimenting with the therapeutic use of exciting gases (particularly nitrous oxide) attracted him to the mysteries of nature as surely as verse attracted Coleridge and his friend Wordsworth. Davy was a natural storyteller; among many gifts he possessed great charisma and fascinated women. Men too: Coleridge practically worshipped him. With Wordsworth, Coleridge invited Davy to help edit the second edition of *Lyrical Ballads* in 1800.[22] Coleridge should also be credited with influencing Davy's view of science. It was he who introduced the chemist to the scientific views of Emmanuel Kant, whose works he discovered on a tour of Germany in 1798. Kant's view that the phenomena of the visible universe were all manifestations of simple forces (a sort of extension of the Uniformitarian principle) was crucial to the development of physics and chemistry in the nineteenth century, in which Davy and his protégé Michael Faraday played such pivotal roles.[23]

A decisive event in his early career, if one can identify such a thing two hundred years on, was Davy's discovery of the work of Alessandro Volta. The Italian's paper on his galvanic pile, the first effective electric battery, which was published in French in 1800, caused a huge stir among natural philosophers. Davy and Beddoes arranged for a large pile of their own to be constructed at the Institute in Bristol and began their first experiments with the electric fluid.

There is no scientific development more Promethean than the harnessing of electricity. In its most sublime and natural form, the bolt of lightning, it is Prometheus's gift to humankind, with its ambivalent promise of power and retribution; it became John Martin's figurative trademark. Electrolysis, the tool which for the first time enabled natural philosophers to break down compounds into their individual elements, earned Humphry Davy the sobriquet 'Chemical Prometheus'.[24] Research into electromagnetism led to the emergence of Michael Faraday as the towering figure of nineteenth-century science, and to the birth of modern physics. But in 1800 the electric fluid was as yet merely an elusive sprite hiding at the bottom of Pandora's jar. In his attempt to capture and tame it Humphry Davy was irresistibly drawn to London, where he was offered the chance to become its master.

CHAPTER FOUR

Mechanics of war

IN NEWCASTLE UPON TYNE the reality of war was setting in. The price of wheat rose dramatically. In 1796 a temporary grain store was erected on the New Road behind Sandgate to hold 120,000 bushels of wheat in reserve. Locals nicknamed it 'Egypt'.[1] At Redheugh, on the north side of the river, the eccentric outline of a shot-tower rose against the backdrop of the Tyne valley. When finished it stood at 175 feet, roughly the height of Nelson's column.[2] Looking like a tall lighthouse with a bulbous head, it accommodated a furnace at the top in which lead was heated until it became molten. The hot lead was poured through a copper mesh and when the globules reached the base of the tower they plunged with a hiss, perfectly spherical, into a pool of water.

Shot-towers were the creation of William Watts, a Bristol plumber. Before his inspired idea, shot for muskets had either been cast in moulds or made by pouring hot lead through a copper sieve into a barrel. Casting was slow and expensive; the barrel method produced teardrop-shaped shot which had to be rounded by hand. The vastly increased demand for munitions during the American War of Independence forced a technological arms race. Watts adapted his three-storey house in Bristol so that he could experiment with the height of the drop and showed that if it was high enough, surface tension would have enough time to act on the lead to result in truly round globules. He was granted a patent in 1782. His own improvised tower in Bristol was still making lead shot as late as 1968 when it was demolished to make way for a road improvement scheme. Shot-towers were a common sight in the cities of Napoleonic Britain, a constant and visible reminder of a state of war that began to seem permanent. Both J. M. W. Turner and John Constable included them in landscapes, confident that their public would appreciate this pointed nod to modernity and war. So far as one can tell, there are now just three surviving towers in Britain: at the Chester Leadworks, at Sheldon Bush in Bristol, and the recently restored example in Crane Park, Twickenham.

Shot-towers and other technical innovations were desperately needed as the war became a deadly serious business. That continental conflict might spread across the Channel was confirmed at the end of 1796 when an abortive invasion fleet carrying Wolfe Tone and fifteen thousand French troops aimed at fomenting an Irish uprising attempted to land in Bantry Bay.[3] It was defeated by a combination of naval vigilance and terrible weather. Further invasion fears in the next year, accompanied by naval mutinies at Spithead and the Nore, saw Newcastle's banks close their doors in a state of near panic.

It was in such a climate in 1798 that Newcastle witnessed the experimental conversion of a keelboat into a gunboat when an 18-pounder cannon was successfully mounted in its bows. Whether riotous keelmen were encouraged to command such vessels is doubtful. There was a very real fear that patriotism might not extend to an urban labouring class already notorious for flexing its industrial muscle.

For a while in 1798 invasion fears eased. Nelson's Mediterranean squadron caught up with Bonaparte's Egyptian invasion fleet at Aboukir Bay, destroying it and stranding the precocious Corsican general. But if optimists saw in this victory the beginning of the end of the revolutionary wars they were wrong. By the end of 1799 Napoleon had escaped from Egypt and effected a *coup d'état* in Paris. As First Consul he became France's de facto dictator. For the next six years he contemplated an invasion of England while attempting to strangle British trade. It took another nine years for Britain to gain her first foothold on continental Europe and seven more after that to bring about final military victory. It was the length and scale of the war – a 22-year global conflict – that cemented Britain's technical revolution.

While in England scientists (Joseph Priestley apart) had been more or less ignored, in France they were actively discouraged during the first phase of the Revolution. Antoine Lavoisier, former tax collector and the founder of modern chemistry, was guillotined in 1794 (his fate may have been sealed by a contretemps with Jean-Paul Marat). Marc Brunel had escaped and could now render his services to the British war effort. The inventor Joseph Jacquard was mobbed by the weavers of Lyon and his innovative loom broken up.

Somehow the great naturalist Georges Cuvier managed to survive, holding appointments under the Ancien Régime, the Revolution and the empire. Perhaps '*les pouvoirs*', the authorities, did not find vertebrate palaeontology threatening (not realizing where it would lead). Perhaps it was Cuvier's phlegmatic nature: he was once famously woken from his sleep by a group of students dressed as devils, who told him they were going to eat him. He took one look at them and said, 'I doubt whether you can. You have horns and hooves; you only eat plants,' and went back to sleep.[4] The irony in the title of his 1812 *Discours sur les révolutions du globe* was no doubt intended.

In Britain there were new opportunities for technical innovation and plenty of talent to draw upon, but as yet no professional class of engineers to direct or nurture it. Clever men still emerged haphazardly from the workshops of their fathers. But that was about to change. Joseph Bramah, a Barnsley farmer's son whose inventive imagination ranged from water closets to hydraulic presses, beer pumps and beyond, spawned no less than three generations of engineers in his London workshop. Unfit, through a leg injury, to follow his father's horse or oxen in its furrow, he was apprenticed to a village carpenter. After his release he was drawn from South Yorkshire to London, where it was said a man might make his fortune. Here he founded the first and most important of the technical professions: precision toolmaking. Along with Josiah Wedgwood in Birmingham he was one of the first to wrestle with the problem of reproducing technical perfection on a profitable scale. Such perfection had existed for decades on the continent, where designers of clocks and automata worked to a precision unmatched anywhere in the age of manual craftsmanship; but it had barely been industrialized. The torturous experience of carpenter-turned-clockmaker John Harrison with his longitude chronometer might have been enough to put any Englishman off.

Settling in London, Bramah set up his own business and began to diversify as his talents drew him to solve some of the capital's pressing technological problems. In 1784 he invented a padlock which stands as an icon of its era. It was constructed with a series of levers and springs of such dazzling complexity and precision that Bramah believed no one would be able to pick it, let alone copy it. He displayed the lock in the window of his shop at 124 Piccadilly with a label that read, 'The artist who can make an instrument that will pick or open this lock shall receive two hundred guineas the moment it is produced.'

This was confidence on the Herculean scale of a George Stephenson or an I. K. Brunel; the brag of a man wearing a big hat to keep the rain off a big ego. But Bramah's promise was not idle and nor was his confidence. Many a cocky young man tried his hand at the lock. None succeeded. The lock, now preserved in the Science Museum in London together with its provocative label, was not opened until it was exhibited 67 years later at the Great Exhibition in 1851, when it was picked by an American locksmith who spent 16 days at the task. There is still some debate about the legitimacy of his method. Bramah had been dead for 37 years but the company, which still makes locks today, duly paid up.[5]

The brilliance of the unpickable lock relied very much on the precision with which its slides, levers, springs and barrel were made. Making money out of such a potentially valuable invention lay not in producing a single lock, however, but in manufacturing it on a large scale. Bramah knew that the answer lay in designing precision tools to reproduce the mechanism to a very fine tolerance. He found his solution at the Woolwich Arsenal, where an 18-year-old called Henry Maudslay was serving an informal apprenticeship in engineering – he had in fact been born in the Arsenal, where his father was a carpenter. Henry preferred the smithy's shop to that of the joiner and by the time of the Revolution in France he had acquired a reputation as a highly skilled metalworker and a young man full of clever ideas for solving problems.

In spite of the resentment of his own apprentices Bramah offered Maudslay a job and set him to work on the problem of lock-making tools. One of Maudslay's most important contributions to machine-toolmaking was the idea of introducing machine tools built entirely of metal – tools that had traditionally been crafted from wood with steel or iron parts where necessary. All-metal tools gave extra stability and precision, and in turn led to the fabrication of yet more accurate tools: it was a self-propelling Promethean revolution. Between them, with Maudslay taking the lead, the machinists in Bramah's works devised a series of mechanical operations which would reproduce the parts of his lock mechanism with perfect accuracy:

> The secret workshops … contained several curious machines, for forming parts of the locks, with a systematic perfection for workmanship, which was at that time unknown in similar mechanical arts. These machines had been constructed by … Mr. Maudslay, with his own hands, whilst he was Bramah's chief workman … The machines before mentioned were adapted for cutting the grooves in the barrel, and the notches in the steel plates … The notches in the keys, and in the steel sliders, were cut by other machines, which had micrometer screws so as to ensure that the notches in each key should tally with the unlocking notches of the sliders …[6]

Maudslay went on to solve another problem that had been exercising inventors for a long time. Bramah had designed a hydraulic press, based on the well-known but counterintuitive principle that a tall but narrow column of liquid can be manually depressed to raise or lower a piston of large diameter a small distance: perfect for pressing, and for raising heavy weights. It is the principle applied in all hydraulics from the car jack to the mechanical excavator. Its effectiveness relies on the use of a tight seal – as all piston/cylinder machines do. Bramah's first attempts at a press, designed principally for the printing trade, could not achieve a tight seal between the piston and the sides of the cylinder.

It was a problem for steam-engine designers too. In the low-pressure engines of Newcomen, Watt and others, materials like hemp rope were used and engineers accepted and allowed for a certain amount of leakage. James Watt believed that partly because of the piston seal problem the development of a high-pressure steam engine was impracticable.

Maudslay's idea was to use a convex leather collar fitted into a recess around the wall of the cylinder.[7] When water was forced into the cylinder at high pressure it flapped out the bent edges of the collar which were forced against the sides of the rising piston, forming a tight seal; when the pressure was released the collar collapsed, allowing the piston to return to its starting position.

Maudslay became Bramah's works manager and stayed with him for more than seven years. Ultimately Maudslay made the mistake of asking Bramah to increase his wages of 33 shillings a week; and Bramah made the mistake of refusing him. Maudslay left and set up on his own in a small smithy and workshop off Oxford Street.

Within a year he had patented the slide-rest, then as now regarded as a fundamental tool of mechanical technology. It was a mechanism for holding a lathe's cutting-tool perfectly steady as it was drawn along the face of the piece to be turned – it might be a piston, a screw or a valve. The rest drew the cutting tool along by means of a threaded bar wound by a handle. Assuming the bar to be perfectly straight (and Maudslay had an obsession with planed surfaces) any turned piece would be absolutely uniform along its length to a very fine tolerance. It could also be reproduced exactly.

The great Scottish engineer James Nasmyth, who became one of Maudslay's celebrated pupils, later wrote:

> It is not, indeed, saying at all too much to state that its influence in improving and extending the use of machinery has been as great as that produced by the improvement of the steam-engine in respect to perfecting manufactures and extending commerce, inasmuch as without the aid of the vast accession to our power of producing perfect mechanisms which it at once supplied, we could never have

worked out into practical and profitable forms the conceptions
of those master minds who, during the last half century, have so
successfully pioneered the way for mankind. To Henry Maudslay
... we are certainly indebted for the slide rest, and, consequently, to
say the least, we are indirectly so for the vast benefits which have
resulted from the introduction of so powerful an agent in perfecting
our machinery and mechanism generally.[8]

Maudslay's first application of this device was to make screws of
uniform thread, hitherto a job for the most highly skilled craftsmen.
Later, the makers of pistons, rods and cylinders realized its great potential for reducing the tolerance of steam engine parts: pistons fitted
cylinders more exactly, increasing their performance and longevity;
valves became smoother and more efficient. The slide-rest, as much
as the iron rail and the high-pressure boiler, was a progenitor of the
locomotive and it led to an even higher level of precision, for with it
Maudslay made a bench micrometer which could measure tolerances
to an extraordinary one ten-thousandth of an inch.[9]

In 1800 or 1801 Maudslay's brilliance came to the attention, quite
by chance, of Marc Brunel. On his arrival in England in 1799 Brunel
had used his letter of introduction to the First Lord of the Admiralty,
Earl Spencer, to interest him in his ideas for mechanizing the block-
making machinery in the naval dockyard at Portsmouth. The consumption of pulley-blocks by the Royal Navy was in the region of one
hundred thousand per annum. Blocks were used in scores of operations on warships wherever mechanical advantage was required. They
consisted of wooden shells within which were mounted one or more
wheels around which ropes could run. Often used in pairs, they enabled
two, three and four times the force of a single pull to be applied to
clews, tacks, braces, shrouds, anchors and any lifting gear required on
a ship. A 74-gun two-decker warship needed four hundred of them,
plus any number of spares. There were shoulder and sister-blocks,
single, double and triple blocks; cat blocks for the anchors, tackle
blocks for the guns and snatch blocks for the fore and main sheets.[10]
Each type of block was different and the problem of mechanizing the

entire process, though not a new idea, was thought to be insurmount-able. Taylors of Southampton, the government's contractor for blocks, thought so. Brunel approached them with his scheme and they refused to consider it. However, a new Inspector-General of the navy works, Samuel Bentham (brother of Jeremy) had been appointed in 1796. He had his own ideas about mechanization, having already installed a 12-horse-power steam engine in the yards. But he looked at Brunel's scheme and, perhaps influenced by Brunel's impressive credentials, recommended it to the Admiralty.

Brunel was a brilliant conceptual engineer. He envisioned a system for manufacturing pulley-blocks that would mechanize the entire process from log to sheave, so that the required output could be achieved by ten men instead of a hundred. But he was not a practical mechanic:

> I received an order to be at the Admiralty with my small models, which gave such satisfaction that my proposition was adopted. Accordingly, General B. took me to Portsmouth. Having had the occasion then of seeing what had already been done by the steam engine and building I made my dispositions accordingly. But the most difficult task was to find some person fit for the execution of so extensive and so complicated an apparatus …[11]

It was a friend of his called Bacquancourt, another French émigré who had passed Henry Maudslay's shop window many times, who put the two men together. Brunel was initially cautious. He did not want to reveal too many of his ideas to a man who might steal them. He showed Maudslay carefully selected drawings of small parts of the machines that he envisaged, asking his advice on their construction. After two or three visits Maudslay, whose father was, after all, a carpenter, is said to have exclaimed, 'Ah! Now I see what you are thinking of; you want machinery for making blocks.'[12]

The process had to begin by roughing out shells using a circular saw. The shell would then be bored, mortised, shaped and scoured to create the block. The sheave or wheel, which rotated inside the block, would be fashioned on a separate sawing machine which simultaneously trimmed the wood and bored a hole through it. Lathes would be used both to finish the sheave and turn the pin on which it rotated.

It was a hugely ambitious project. Even with Maudslay's energy and skill it was not until 1805 that the machines, more than 40 of them, each carrying out a single process, were all up and running. They are at once exactly fit for the purpose and aesthetically delightful; and it is fitting that they now sit within yards of Joseph Bramah's unpickable lock and Richard Arkwright's water frame in the Science Museum in South Kensington. Some of Maudslay's block-making machines look like the scaled-up innards of a carriage clock, others like the guts of a jet turbine engine. There is not an ounce of superfluous metal; everything is perfect, and perfectly in its place. As Maudslay often intoned to his workers, the mechanic must ask of his design, 'What business has it to be there? Avoid complexities,' he would say, 'and make everything as simple as possible.'[13] From crosscut saws to mortisers and shapers, these devices were virtually unimprovable. Some of them were still operating in the middle of the twentieth century, which is why several have been preserved. One of the most elegant concepts in the manufactory was that marks made by the tool gripping the roughed-out shells were used to locate the block exactly on corresponding pins in each subsequent machine. It appealed to Maudslay's fixation with accuracy, and it appealed to Brunel's love of an elegant concept.

Brunel may have made a major contribution to the war effort but he was neither the first nor the last contractor to find that the British government's gratitude was more theoretical than financial. He had to wrangle with them for a cut of the savings made by the dockyard, and in the end made seventeen thousand pounds, the amount the government thought it was saving every year; scant reward for six years of his life. Brunel seemed destined to become one of those who made more money for other people than for himself; he later suffered the ignominy of a spell in the King's Bench debtors' prison. There were plenty of men like that throughout the period of the Industrial Revolution. It seems also to have been a particular feature of government contracts, reminding one again of John Harrison's struggle to claim his reward for the longitude chronometer. Both William Martin and Charles Babbage, the pioneer of the digital computer, spent their later years complaining that their talents had not been sufficiently rewarded.

The next generation of engineers (and in this period the same goes for writers, artists and scientists), the true professionals, were much cannier about making money. Those older visionaries, Marc Brunel, William Martin, George Stephenson, Richard Trevithick: the intuitive, untrained, empirical 'amateurs', did not, or did not want to, understand what came to be known as 'business'. Their sons, younger brothers, apprentices – John Martin, Isambard Kingdom Brunel, Robert Stephenson – were of a new stamp: rational, deductive, professional. They learned technical skills from their parents and masters, and business from their competitors.

CHAPTER FIVE

Brothers in arms

THE BUSINESS OF WAR and its technological payoff was not fully felt in Britain until the army set foot on Continental Europe after 1808. In the long term it brought the country economic and strategic dominance over much of the world during the nineteenth century. Before that it produced two decades of domestic turmoil in the aftermath of Waterloo.

The war emasculated opposition to successive Tory governments and suffocated the reform movement. At the same time it created a theatre of opportunity. In the brief lull provided by the Peace of Amiens in 1802 men and women of intellect and enterprise crossed the English Channel in their thousands. Among visitors to Paris were William Wordsworth, J. M. W. Turner, the novelist Maria Edgeworth, the critic William Hazlitt, reformer Sam Romilly and no less a radical than Charles James Fox, who dined with Talleyrand, Bonaparte's brilliant Foreign Minister. Natural philosophers caught up with each other's work in a spirit of fraternal cooperation that was only a little dimmed by the renewal of hostilities in 1803.

Many of those who, like William Martin, had served in the militias, now returned to their families or their work. There was no shortage of employment in the northeast of England where new collieries were won at Percy Main and at Jarrow on the south bank of the Tyne. The Northumberland Militia was disbanded in April 1802 amidst a general, if illusory, sense of joy and relief: 'On our arrival at Gateshead Fell, we were met by a multitude of people to welcome us home again, and the night we approached the town, the crowd increased so much that we could scarcely reach the Tyne Bridge.'[1]

The peace was a sham, as kings, ministers and dictators knew. While an impoverished British government was busy reducing its navy, Napoleon Bonaparte embarked on a substantial shipbuilding programme, began planning his invasion of England and blocked all British trade with European ports that lay under his control or influence. The war resumed a year later with increased determination on both sides to annihilate the enemy. The England of 1803–5 was almost obsessed by the threat of invasion. The much-maligned Duke of York, Commander-in-Chief of the army, rather sensibly organized a system of homeland defence that dissuaded Napoleon's Army of England from embarking some months before the moral victory at Trafalgar. It included Martello towers built at dozens of strategic points along the south coast; the Royal Military Canal at Rye in Sussex, and a series of rapidly deployable redoubts on the Downs overlooking London. Nelson was given charge of the navy's operations in the Channel to harry the French invasion fleet in an exercise designed more to calm domestic panic than to achieve a military objective. Many of the militia regiments disbanded in 1802 were re-formed.

William Martin, after visiting his parents, did not rejoin the militia but instead returned to the Howdon ropery where his fertile imagination sired designs for a paddle boat, a harpoon for whalers, shoes 'on a new principle' and a pneumatic lifejacket for saving lives at sea.[2] Richard Martin, having returned from a long spell in Ireland, remained in the army. In 1805 he was drilling volunteers at North Shields but he took time off to visit the local Member of Parliament with William to show him his brother's plans (with drawings made by John) for a mining fan ventilator. Richard then joined what later became the Grenadier Guards (until 1816 the 1st Regiment of Foot Guards). Rising to the rank of quartermaster-sergeant, he fought with them in Spain and Portugal under Wellington, but because the regiment has no personnel records for this period he becomes invisible once again.

John Martin was too young to become involved in the war. He had already, he later wrote, decided to pursue painting as a career.[3] It does his parents great credit that they not only failed to dissuade their son from such a career but actively encouraged him. His talents may have prompted the Martins' move into Newcastle in 1803. John, aged 13, was apprenticed for seven years to a coachbuilder called Leonard Wilson. On completion of his apprenticeship it was expected that he would become a journeyman in the lucrative trade of herald painting which exploited the bourgeois fashion for setting up one's coach and having it adorned with a coat of arms, authentic or not.

John may have been driven by vocation but he shared his family's restless genes and came to realize that the drudgery and humility required of an apprentice (14-hour days were not unusual) were not for him. He began to resent his master, the tedium and the poor remuneration. He was lucky in a way: his master provoked him into rashness. The articles under which he was bound apprentice, his indentures, stated that after his first year he was to be given a rise in wages. Wilson's habit was apparently to claim that the first three months of the year counted as a trial, so the wage-rise was deferred. Other apprentices put up with such chicanery if they knew what was good for them. When his year was up John duly demanded his rise, was refused, and confronted Wilson. 'I won't submit,' he told his master, and walked out.[4]

The Newcastle newspapers of that time were full of advertisements offering rewards for capturing absconded apprentices in their *Hue & Cry* banners. In what one might call the Dickensian scenario of 40 years later John would have been thrashed by his unsympathetic father, thrown into a coal cellar and half-starved on thin gruel to bring him to a proper sense of his duty. But when John got home after walking out on Wilson, Fenwick Martin congratulated him on sticking up for himself and gave him a shilling to go and buy drawing materials. John duly spent the next few days with a set of new pencils at his bedroom window making studies of the sky, especially of lightning, and keeping a low profile. When the bolt fell, it was not lightning but the Town Sergeant that appeared in his room and dragged him off to appear before the Guildhall Court.

In this age of surprises, of extremes of tolerance and intolerance, the unexpected must be expected. The Beadle turned out to be a rather decent man. He stopped to admire John's pictures before escorting him away. John's parents being out at the time of his arrest, he panicked and tried to escape but the Beadle was reassuring and persuaded him to come quietly. At the court, with Wilson in the witness stand, John was charged with violence, having run away, rebellious conduct, and threatening to do a private injury. But to Wilson's disgust the witnesses he brought to support his case corroborated John's version of events and the 14-year-old was acquitted. He then asked for his indentures to be returned to him. The Alderman ordered that it should be so. According to Martin, writing 40 years later and relishing his own precociousness, 'I was so overjoyed that, without waiting any longer, I bowed and thanked the Court, and, running off to the coach factory, flourished my indentures over my head crying: "I've got my indentures and your master has taken a false oath; and I don't know whether he is not in the pillory by this time!"' [5]

John, the most self-possessed and rational of the Martin brothers, could not resist kicking against the pricks. So it is surprising that Jonathan, the least stable, was able to stick at his tannery apprenticeship until the age of 22. But when he was released from his indentures in Hexham in 1804 he decided to travel abroad. It was perhaps not the best year to embark on a tour, grand or otherwise. The continent of Europe lay prostrate under the boot of the newly self-crowned Emperor Napoleon I. In the Atlantic, the Mediterranean, the Caribbean, the Indian Ocean and the Baltic Sea the Royal Navy pitted its overstretched resources against France and her allies. The world was at war.

Jonathan might have followed in the footsteps of explorers like Mungo Park, soon to embark on his second trip to explore West Africa (he was killed there in 1806). But such dangers did not suit a man as timid as Jonathan. He might have looked for adventure closer to home. In Scotland Thomas Telford had begun to construct his technically ambitious Caledonian Canal linking the Atlantic with the North Sea, and at Penydarren in Wales Richard Trevithick was demonstrating his second experimental locomotive. Instead, Jonathan made his way to

London as Joseph Bramah had done although, in Jonathan's case, with no very clear idea of what he might do.

Then as now, naïve young men arriving on the streets of the capital for the first time were easy prey for sharps. It is generally supposed that Jonathan fell victim to a press gang, but this cannot have been quite the case. Even in time of war the press operated under strict rules. Only men who had already been to sea could be 'impress'd'. They were spotted easily enough by their rolling gait, trousers, tattoos and pigtails, and by 1804 there can have been very few of them left at liberty on the streets in London or anywhere else. Britain had mobilized a hundred thousand of them and they now found themselves enduring the hardships and ample diet of the navy on blockade off Ushant, Toulon or Cadiz. If they wished, landsmen could volunteer to serve at sea, for which they were paid a bounty. If they did not, they might still fall into the hands of unscrupulous 'crimpers'. These were the sort of men against whom the glass bottoms in pewter tankards were designed so that innocents like Jonathan should not accidentally take the King's shilling with their ale or porter.

In Jonathan's case it was rather simpler. He later recalled, 'One day while viewing the Monument a man accosted me (perceiving that I was a stranger in Town) and enquired if I wanted a situation, I informed him of my desire to go abroad: he said he could suit me exactly ...'[6] The man offered him 32 shillings a month plus prize money to go to sea as a temporary replacement for the son of a gentleman of the man's acquaintance. This seemed like an ideal opportunity for a brief adventure. But Jonathan soon found himself confined in a miserable lodging with several other men and a bolt on the outside of the door. He was sworn in (hardly stretching the limits of his imagination, he gave them the false Christian name John) and taken aboard the *Enterprise* schooner which lay on the Thames. From there, he wrote, he was transferred to the third-rate 74-gun *Hercules*[7] lying anchored at the Nore, where he began his inadvertent naval career far from the comforts and security of his family and the green valleys of Tynedale.

Jonathan's colourful account of his time in the navy needs to be treated with some scepticism, though in its basics it rings true. He cannot, to begin with, have joined *Hercules* at the Nore in 1804, for she was stationed in the West Indies at the time. In recalling this period nearly twenty years later he probably conflated experiences in several ships. He also claimed to have been made a foretopman on his arrival in *Hercules*, but whatever ship he did initially serve on, as a landsman he is most unlikely to have been sent to the foretop so quickly. This was where the most skilful and agile seamen were usually to be found, so Jonathan must have learned less demanding ropes to begin with. According to his own testimony, he also prayed a great deal.

That Jonathan was, as he claimed, serving in HMS *Hercules* at the siege of Copenhagen in autumn 1807 there is no doubt, because at his famous trial in 1829 the prosecution found an old shoemaker who had sailed with him. By the year 1807 Jonathan had indeed been made captain of the foretop, from where the shoemaker remembered him falling directly into the sea. As with all the events in his life, Jonathan saw this providential escape as a sign from God. He had three other lucky escapes after falls from the yards or rigging. Such accidents were accepted hazards of naval life, but to remain uninjured after so many scrapes was indeed providential. His brother William, who like the rest of the family had a strong affection for Jonathan, later made a crude woodcut, which survives, celebrating these events.

In 1808 *Hercules* was retired and Jonathan joined another, unnamed, ship bound for Lisbon to blockade the Russian squadron lying in the Tagus. Thus both he and his brother Richard were present at the start of the Peninsular War. Napoleon's fatal error in provoking the Spanish uprising of the Dos de Mayo (celebrated in Francisco de Goya's famous paintings) was quickly capitalized upon by Admiral Collingwood's [8] Mediterranean fleet and by an expeditionary force under Sir Arthur Wellesley. It was the army's first chance to engage Napoleon on Continental Europe. After a particularly bloody and unattractive campaign lasting seven years it led to the Emperor's nemesis and his exquisitely Promethean exile tethered upon a rock in the South Atlantic. [9]

After a brief return to Portsmouth (there is a ludicrous tale about Jonathan deserting his ship here to buy a box of paints and being forgiven by a kindly lieutenant; so ludicrous it may even have happened) news reached the squadron of Sir John Moore's disastrous but heroic retreat to La Coruña.

After his heroics at Rolica and Vimeiro, when he drove the French from Portugal, Wellesley had been withdrawn and replaced by Sir Hew Dalrymple. At Cintra in August 1808 Dalrymple signed a dishonourable treaty with the French, who were allowed to withdraw with all their equipment in British ships. Dalrymple was recalled and court-martialled amid a huge public outcry. In October Sir John Moore was sent to the Peninsula with a second force which found itself outnumbered by Marshal Soult's army and was forced to retreat to La Coruña. From here, nineteen thousand men were brought off by the navy but with the loss of their commander, fatally injured during the rearguard action. Wellesley, soon to become Viscount Wellington, was restored to the command and did not return to England until the war was won in 1815.

Jonathan's eyewitness portrayal of this first 'Dunkirk' rings only too true:

> By great exertion the whole embarkation was completed. They [the French] then directed their batteries against our transports, who had to slip their cables and then stand out of the reach of their guns. During this scene of confusion and terror several boats were sunk by the fire from the enemy, and some by the violence of the sea. Our vessels presented an awful spectacle from the number and condition of the wounded, who occupied our cockpit, cable tier, and every spare place on board; and whose misery was rendered greater by the tempest which arose ... a great number perished ...[10]

There is very little so far in Jonathan's account of his time in the navy to hint at his complex and fragile mental state. He claimed to have been proud of his service and his sailor's skills, although he thought the better pay and easier discipline of the merchant service might suit him more. He was the least martial of the Martin brothers. The first signs of instability show in the account of his multiple desertions. Initially these came about because he was sent back to Lisbon with a prize crew in charge of a Spanish vessel of dubious neutrality. Sent ashore to collect brushwood, Jonathan and some of his mates decided, apparently on the spur of the moment, to desert to the enemy. Ill-prepared for their adventure they got drunk at a wine-house and Jonathan was somehow relieved of his trousers. After a couple of days of aimless and thirsty wandering they found themselves back within sight of their ship on the Tagus. In Jonathan's account they decided to throw themselves on the mercy of their captain (probably in fact a second or third lieutenant), were pardoned, and returned to the ship where Jonathan thought long and hard on his own wickedness.

This did not stop him from making another bid for freedom, this time further up the Tagus where they were to land troops from shallow-draughted barges. Now a veteran escapee, Jonathan thought to prepare a parcel of food and took himself off alone. In a passage as implausible as anything from *Gulliver's Travels* Jonathan recounted how he had fallen in with the Portuguese corn harvest, had joined in, been taken into a farmer's house and somehow introduced to an eligible young lady with whom he might settle and produce children. All of which sounds too good to be true, and probably was. Jonathan came to fear that the temptations of this attractive opportunity were a test set by God: the Portuguese were, after all, Catholic heretics. Again defying sense and belief, he returned to his ship, pleaded for his officers' mercy, and was pardoned. Whether the pardon was accompanied by a few dozen penitential lashes he does not say.

That Jonathan had guts there is no doubt. Now returned to a ship of the line sailing from Lisbon to Cadiz, he was assigned to a gunner's crew. At some point in the voyage a careless gunner's yeoman managed to shoot himself in the head in the powder magazine. Amid the smoke and subsequent confusion the fear of explosion panicked the crew, some of whom attempted to abandon ship. Jonathan, with four other gunners' mates, went into the smoking magazine, extinguished what turned out to be a small fire caused by burning wadding from the yeoman's pistol, and found the victim lying fatally wounded on the floor. This extraordinary act of bravery was later corroborated by a writer who managed somehow to track down a number of his shipmates from this period.[11]

Jonathan claimed it was frequent gunfire from the shore batteries of Cadiz that determined him to desert a third and final time. Quite what the batteries of Cadiz were doing firing on an English ship after 1808 when the two countries were allies is hard to fathom: it sounds like one more of Jonathan's embroideries. What is certain is that he did not return to his ship after his third desertion, and was never caught. Instead he managed to find passage on board a transport heading for Egypt to fetch corn for the army.

His shipmates' accounts of his time in the navy offer much more than mere confirmation of the main points of Jonathan's story. They also reveal his paradoxical character: at one time sober, reliable and expert, at another boring his mates rigid with biblical cant, drinking and deserting with them, making up stories and lying when he had done wrong. He was thought to have been a shirker and a hypochondriac by some; others thought him 'soft'. He would talk at great length about celestial wonders but had a superstitious abhorrence of anyone pointing at a star. He denied the existence of other worlds. Others said they had heard him talk often of dying, and of a future state. He was sometimes ridiculed for his extreme beliefs and this sent him into fits of melancholy.

Having lost one son to the navy in 1804 and with Richard also having served abroad, Fenwick and Isabella Martin were keen that their youngest surviving son John should stay close to them in Newcastle. The 15-year-old, after breaking his indentures, would not easily find another apprenticeship in the northeast, so for a time John was happily free to wander, as his father was wont to do, among the hills and woods of the Tyne valley. But he was painfully aware that his attempts to draw and paint the sublime beauties of nature were no more than amateurish daubs. His only professional satisfaction, if that is the right word, came when William asked him to draw up plans of his mechanical inventions. Apart from the fan ventilator scheme presented to the local MP by Richard, William also had an idea for a means by which coal might be carried more efficiently from pit-head to screen than by horse-drawn wagon. According to William the plan was stolen from him (this is to become a familiar refrain) and none of the brothers had any benefit from it.

This is a striking point in the Martins' story in which one traces the passing of an empirical, undisciplined generation. William's talent for envisioning inventions and new processes was real. But he had no formal training because there was none to be had outside colliery smithies or the workshops of wheelwrights – none except perhaps in London where Bramah, Maudslay and Brunel were forging their new profession. William lacked the opportunity but also the will to undergo the rigours of training and set himself up as a mining engineer. Amateurism in fact sat comfortably with him, frustrated though he ultimately was. Perhaps it suited him, like Marat, to trade on a strong sense of injustice. It was one thing to suggest grand schemes of improvement; quite another to set about building them.

A contemporary of Jonathan's, George Stephenson, was the man for that. Born in Wylam on the River Tyne in 1781, the son of the colliery engineer there, he had the upbringing and overwhelming desire to succeed as both engineer and man of business. This is not to say that he is as attractive a figure as William; but he knew how to get things done. At the time of William's Howdon inventions Stephenson had left home illiterate, taught himself to read at night school, married, and become engine-man at Willington Colliery at the age of 21. Here he began to use his empirical knowledge to improve the steam engines under his care. Later, as the army's consumption of horses raised their cost to prohibitive levels, he turned his attention to the locomotives being developed by Trevithick, Blenkinsop and Hedley. It was later said of his engines that what was good was not original and what was original was not good. He himself propagated the history that hails him as the father of the locomotive and in reality he was no such thing. But as what we might now call a project manager or industrial entrepreneur he was an undisputed, if unscrupulous master.

For John Martin too an amateur interest in art was not enough. He had ambition and, more importantly, the determination to acquire the skills he needed despite his early setback with Wilson the coach-builder. So now he begged his father to find him professional instruction:

My joy was indescribable when I succeeded in persuading my
parents to afford me a little education ... My father, thinking well

of my talents and the progress I had made without aid, determined
to procure me a master. He went one morning in search of one,
and returned apparently much elated, saying: 'Come with me,
John, for I have found a gentleman who can teach you everything.'[12]

The gentleman in question was an exiled Lombardy artist, Boniface
Musso, who lived in Newcastle and taught both art and swordsmanship.
His skill with the blade may have been the reason that he came to
Fenwick's notice. Musso was extremely kind to John. When Fenwick's
money ran out after the first quarter's tuition, Musso agreed to keep
John as a pupil for free. Now he began his education in earnest; not
just in technique, but in learning the works of the eighteenth-century
masters Claude Lorraine, Salvatore Rosa and Stefano della Bella, of
whose works Musso had engravings. The twice-weekly drawing lessons
became daily and Musso also gave him instruction in oil painting on
Sundays. He formed a firm and lasting friendship with Musso's son
Charles. It is possible that in Musso's collection was an engraving of
Rosa's *Prometheus Attacked by the Eagle*, of about 1650; but whatever
his inspiration, John's Promethean insolence had so far paid off.

John began to earn small amounts of money painting portraits on
commission. As his skill grew during this first year he was offered 25
shillings by a widow to paint her son, about to join the merchant navy.
The portrait was artistically successful, at least as far as John and his
tutor were concerned; but he had faithfully reproduced the sitter's
pronounced squint and the work was rejected by the boy's appalled
mother. John would have to make his mind up whether to please
himself or his critics.

After a year this convenient arrangement ended. Charles Muss[13]
was offered a lucrative position at a glass-painting establishment in
London and called his family to move there with him. To John's delight
and his parents' disappointment (especially given that the Mussos were
staunch Catholics) John was asked to join them with the promise of
employment in the glass-painting business. So in September 1806 he
took passage on a coaster bound for London. He was 17.

CHAPTER SIX

A million fires

LONDON IN THE FIRST DECADE of the nineteenth century was the crucible in which the arts of men and women were forged into something recognizable as the modern world, in full view of an ambivalent, reactionary establishment which chose to ignore the obvious. The sharpest of eyes looked on, relishing the irony as the Prometheans set about stealing the future from their betters.

At the end of that decade Sir Richard Phillips (a canny, some said rapacious publisher and patron of the arts who had once been imprisoned in Leicester gaol for selling Tom Paine's *The Rights of Man*) recorded a series of morning walks from London to Kew.[1] His acuteness of observation and sensitivity to the paradoxes of his age make him a cultural historian of a high order. He sets, to begin with, a delightful pastoral scene: the gatekeepers of St James's Park selling the milk of the park's cows to mothers and nurses from sunrise until ten o'clock. In the same park he passes the Prince of Wales' new palace, and bludgeons it aside:

> The love of shew in princes, and persons of authority, is often justified by the alleged necessity of imposing on the vulgar; but I doubt whether any species of imposition really produces the effect which the pomp of power is so willing to ascribe to it, as an excuse for its own indulgences.[2]

From St James's Park Phillips proceeds to The Mall, where he pauses long enough to sketch the stratified emergence of the nation's wealth-creators. Before nine in the morning the clerks are about, walking with a busy sense of purpose; from 9 to 11 come the shopkeepers, stock-brokers and lawyers. Then,

> At twelve saunters forth the man of wealth and ease, going to look at his balances, orders or remittances; or merely to read the papers and hear the news; yet demonstrating the folly of his wealth by his gouty legs, or cautious rheumatic step.[3]

Phillips was not merely a satirist. As befitted an age when a man might still know all that was to be known and in which all things seemed to be connected, he was intensely interested in all things. He took in, opposite the Pimlico waterworks, 'the manufactory of the ingenious Bramah, whose locks baffle knavery', and when he arrived at the river near Battersea Bridge he stopped to admire 'the workshops of that eminent, modest, and persevering mechanic Mr Brunel; a gentleman of the rarest genius, who has effected as much for the Mechanical Arts as any man of his time'.

The workshops in question were Marc Brunel's sawmills, one of the wonders of the age. Brunel, not having made as much money from the Portsmouth block-making manufactory as he could have wished, had installed at Battersea a 16-horsepower steam engine to run three enormous circular saws, one of them 18 feet in diameter. These cut veneers of as little as a sixteenth of an inch thick. The operation was run for the most part by children, a fact which acquired for Brunel the dubious reputation of being a philanthropist. Phillips, impressed as he was, saw at once the other side of the coin: 'owing to social monopolies, and to the advantages taken of poverty by wealth, the mass of the people are less benefited by the introduction of machinery than they ought to be', he wrote.[4]

Phillips did not fail to pay his own sort of homage to London's river. Blake had associated the Thames with 'marks of weakness, marks of woe'.[5] Wordsworth, in London during the Peace of Amiens in 1802, had had more romantic thoughts 'Upon Westminster Bridge' ...

Earth has not anything to show more fair:
Dull would he be of soul who could pass by
A sight so touching in its majesty:
This City now doth, like a garment, wear
The beauty of the morning; silent, bare,
Ships, towers, domes, theatres, and temples lie
Open unto the fields, and to the sky;
All bright and glittering in the smokeless air …

Phillips' unromantic brilliance was to observe minutely whilst standing back. Eagle-like, he saw both the detail and the larger picture. And on Putney Heath, looking back across the river, he saw a London that was both magnificent and ailing:

The Smoke of nearly a million of coal fires, issuing from the two hundred thousand houses which compose London and its vicinity, had been carried in a compact mass … Half a million of chimneys, each vomiting a bushel of smoke per second, had been disgorging themselves for at least six hours of the passing day, and they now produced a sombre tinge. Other phenomena are produced by its union with fogs, rendering them nearly opaque, and shutting out the light of the sun; it blackens the mud of the streets by its deposit of tar, while the unctuous mixture renders the foot-pavement slippery; and it produces a solemn gloom whenever a sudden change of wind returns over the town the volume that was previously on its passage into the country.[6]

Phillips' London was not only the wonder of the age; it was also William Cobbett's Great Wen, the London of pickpockets, prostitution and beggary. His Thames was that of Turner rather than Wordsworth.

Joseph Mallord William Turner was gravitating irresistibly towards the water from the homely bustle of his native Covent Garden. The son of a barber and a violently unstable mother, Turner was born in 1775, an almost exact contemporary of John Constable and Jane Austen. Although his family had come from the West Country, Turner himself was a Londoner of the streets, cockney accent and all. As a child he was sent to stay with relatives in the rural riverside village of Brentford; the Thames was his playground and highway. From Twick-enham to the Pool of London he knew the river, its people and boats intimately; he built his own craft too, sailing it as far as the estuary to sketch and paint in watercolours.

Turner was a prodigy. Taken on as a pupil at the Royal Academy School at Somerset House at an early age, he was elected an Associate by the end of 1799 when he was only 24. His brilliance had already been recognized in his early oils of moonlight on the Thames and sunsets over the seashore at Brighton. Despite his unprepossessing appearance – short, with a hooked nose, bandy legs and overly large feet, he was in much demand at 'evenings' and grand dinners. He was a reluctant, taciturn, sometimes rude guest. As success came he was able to set up his own studio and gallery in the pleasant surroundings of Harley Street, where his father became his devoted assistant (his mother having by now been confined in an asylum). Hating to part with any of his works he lived among the clutter of his canvases and drawings. Much later, when John Martin knew him, he lived an enigmatic double life, keeping one house in fashionable Cavendish Square (where Henry Maudslay had had a workshop employing 80 men) and another, 'a squalid place' on the river at Chelsea.[7] 'Here you see my study,' he would say, 'sky and water.'[8] Like Martin, he surrounded himself with scientists. He became a close friend of Mary Somerville, of Michael Faraday and Charles Babbage. He took a close interest in the chemistry of colour-making which, along with a passion for fishing, brought him into contact with Humphry Davy. He was a frequent visitor to the Royal Society, located in Somerset House next door to the Royal Academy and the Society of Antiquaries. His house in Queen Anne Street boasted a water-closet (probably one of those designed by Joseph Bramah) and he shuffled through the streets of London carrying an umbrella that concealed a dagger in its handle.[9]

At the very end of 1805 Turner travelled down to the mouth of the Thames and was given permission to spend time on board HMS *Victory*, returned much battered from Cape Trafalgar with the pickled corpse of England's saviour, Horatio Nelson. Turner made many sketches and took notes of interviews with her sailors for a major work he planned (quite unlike his later theatrical rendition of the battle, commissioned by George IV in 1824) to depict the intimate human drama and horror of *Trafalgar as seen from the Mizzen Starboard Shrouds of the Victory*. This project established his credentials as an artist of not merely technical brilliance, but also of great journalistic sensitivity.[10]

On 8 January 1806 the funeral barge of Lord Nelson passed up the Thames from Greenwich to St Paul's Cathedral, where the Admiral was buried in state. Muffled peals of bells were rung in towns across the country. Looking back on that year it must have seemed like the end of an era; for not only had Nelson gone but William Pitt was dead at 46 before the end of January and his age-old political adversary, Charles James Fox, also died within a few months. It was to a London reeling from these shocks that John Martin came in the autumn of 1806.

John was 17. He had 'a good outfit [and] small though sufficient funds for immediate purposes, notwithstanding my having been robbed of all my loose cash by a poor passenger in the ship'.[11] His hopes were understandably high, but they were immediately dashed. When he finally arrived at the Mussos' house in Cock Court after having got lost on the way from the port, he found there was no work for him – nor for Charles Muss. The letter telling him not to come had arrived in Newcastle too late.

Instead of returning home to Northumberland, John decided to stay in London and chance his arm. He and Charles spent the next months looking for employment while he worked up drawings and sketches of Northumberland from memory and tried to hawk them in the streets. The Muss family put him up (he is said to have slept in the bedroom of the late radical John Wilkes) until such time as he could afford to contribute to his board.

The first work that he was able to sell went fortuitously to a German called Rudolf Ackermann, the great, perhaps greatest, publisher of illustrated books. At the time of their first meeting Ackermann, son of a Saxony coachbuilder, having set himself up in business on the Strand, was preparing for publication his *Microcosm of London*, illustrated with aquatints by Rowlandson and Pugin. Ackermann is today revered not only by antique print collectors but also by automobile engineers, for he designed the first non-parallel steering mechanism which allowed coaches, and later cars, to steer round corners without sliding. He was entrusted with the construction of Nelson's funeral car. He was also a pioneer in the use of gaslighting, having all his printeries and galleries lit by gas from about 1812. The *Practical Treatise on Gas-lighting* by Friedrich Accum that Ackermann published in 1815 is regarded as one of the most beautiful books ever produced in England.[12]

At first Ackermann refused to look at John Martin's drawings, already having more second-rate landscapes in his store than he could sell. But he changed his mind, bought three ink sketches for 12 shillings and told Martin he would take more. Martin rather rudely told Ackermann he would never sell him another work – he had valued his sketches at a guinea and loathed the idea of being exploited. When he got home Charles Muss told him he was a fool. Humility was not a natural Martinian trait. It did not appear in the family genes and experience had already taught John that humility and servility were bedfellows. Years later he found Ackermann a reasonable and supportive publisher of his engravings and was so embarrassed at his early arrogance that he hadn't the nerve to remind the German of their first encounter.

After many lean months Charles Muss went into partnership in a glass and china business and was able to offer John regular work. He was paid two pounds a week while serving an informal second apprenticeship, for which privilege he gave half his wages back. At night, he later recalled, he spent his hours learning perspective and architecture, and eventually began to earn enough to take his own lodgings in Adam Street, Cumberland Place. Although Charles was a much more amenable employer than Wilson the coachbuilder had been, there seems to have been a kind of falling out with the Muss household which John called 'some little differences' – but it was not serious enough to affect their friendship or mutual esteem.

In the years after 1806 other young Promethean talents were trying to make their way in London, with mixed success. Fourteen-year-old Michael Faraday, son of a Sandemanian[13] blacksmith, had just begun an apprenticeship with G. Ribeau, a bookbinder and bookseller on Blandford Street. With his master's consent and encouragement Faraday used the shop as a personal library, reading anything and everything.

The late-developing John Constable, born in 1776, had held a first exhibition at the Royal Academy and returned to Hampstead from a tour of the Lake District thoroughly depressed by the awful weather there. John Keats was still at school, as was Percy Shelley. Charles Babbage, the same age as Faraday, was at school too and oddly enough in Enfield, like Keats. One of his friends there was the young Frederick Marryat, later midshipman under Lord Cochrane, captain in the Royal Navy and popular author of seafaring romps and other yarns like *The Children of the New Forest*. In 1812 Babbage, whose father was a banker of comfortable means, went up to Cambridge where his mathematical brilliance was recognized but where he failed to complete his studies, having more interesting things on his mind.

Humphry Davy had been in London for five years and already showed substantial form as a thief of celestial fire. He had been invited to come to the capital by Benjamin Thompson, one of the founders along with Sir Joseph Banks of the Royal Institution. Dr Beddoes very decently released Davy from his position in Bristol, realizing that a lectureship at the new Royal Institution was too good a chance to pass up. Davy's friends Wordsworth and Coleridge (returned in a panic from his European travels convinced that Bonaparte's agents were after him), themselves periodically in London, might have thought that a good poet was going to waste.

Benjamin Thompson (later Count Rumford) was a most colourful character. Born in America, he married a rich New England heiress and sided with the British during the War of Independence, making himself deeply unpopular with his compatriots. He abandoned his wife for ever, fleeing to London and establishing himself as something of a natural philosopher with a reputation for philandering and a love of money. He experimented with gunpowder and friction after moving to Bavaria in 1785 and there produced a landmark work on the theory of latent heat.[14] After that he divided his time between London and Munich. His inventions included various stoves and ranges, notably the Rumford fireplace which is the prototype of the modern domestic hearth. He also developed the Baked Alaska and the first thermal underwear designed on scientific principles. In London he was much caricatured for his extravagant lifestyle, in particular after the death of his wife when he married the widow of guillotined scientist Antoine Lavoisier: 'Married; in Paris, Count Rumford to the widow of Lavoisier; by which nuptial experiment he obtains a fortune of 8,000 pounds per annum – the most effective of all the Rumfordizing projects for keeping a house warm.'[15]

As a patron of budding scientists Rumford deserves some credit for recruiting Davy as the second professional 'experimental philosopher' in Britain after the Astronomer Royal. It was Rumford's aim that the Royal Institution should guide and educate artisans, but Davy's rapid celebrity soon deflected this purpose. After his arrival at the Institution Davy plunged himself into both his work as its first lecturer and into a new, exciting social scene. He joined a club called the Tepidarians at Old Slaughter's coffee house on St Martin's Lane, perhaps introduced by Coleridge who had written to William Godwin about him. His chemical zeal, together with the republican politics of his new friends (though he himself was anything but a republican) created a fissile milieu. His charismatic delivery and physical beauty ensured that his lectures were soon among the must-see events of London society. When he became ill, in 1807, such was the press of carriages belonging to fashionable ladies outside the Institution's grand façade in Albemarle Street that hourly bulletins on Davy's progress had to be chalked up on a board at the entrance.[16]

That Davy was aware of, and intended to exploit, his own popularity and the Promethean potential of science is apparent from the beginning. Describing the role of chemistry in an early lecture which he had printed by public demand, he made particular reference to metallurgy:

The working of metals is a branch of technical chemistry; and it would be a sublime though a difficult task to ascertain the effects of this art upon the progress of the human mind. It has afforded to man the powers of defence against savage animals; it has enabled him to cultivate the ground, to build houses, cities, and ships, and to model much of the surface of the earth after his own imaginations of beauty. It has furnished instruments connected not only with his sublime enjoyments, but likewise with his crimes and his miseries; it has enabled him to oppress and destroy, to conquer and protect.[17]

He might just as well have been announcing the splitting of the atom a hundred years later. No wonder they called him the Chemical Prometheus.[18] By 1806 Humphry Davy was one of the most famous men in London. In that year he delivered a landmark lecture on electrochemical analysis and followed it in the next two breathless years with the 'discoveries' of potassium, sodium, calcium, barium, strontium and magnesium. The *Edinburgh Review* compared him to Isaac Newton.

Davy's dazzling progress towards understanding some of the fundamentals of modern science was now broadcast to a wider audience by the writings of Jane Marcet, who published in 1806 *Conversations in Chemistry* based on his lectures and on the works of Lavoisier and Cavendish. This one work alone (among many in a long writing career) went to 16 editions in Britain, the same number in America and several in France. Among its avid readers, in Ribeau's bookshop on Blandford Street, a few hundred yards from the Institution, was Michael Faraday.

Davy was not the only brilliant chemist working in London at the time. One of those on whom he relied in his first years at the Royal Institution was Friedrich Accum, an exact contemporary of Brunel, Wellington and Bonaparte born in 1769 not far from Hanover in Germany. After moving to London in the early 1790s Accum came to know William Nicholson, an amateur chemist and publisher of a magazine which became known as *Nicholson's Journal*. Its scientific bent attracted articles by Accum, by Henry Cavendish, John Dalton, Benjamin Thompson and eventually Davy himself. In 1800 Accum set up a laboratory in Old Compton Street, Soho and began to sell scientific apparatus. A year later Davy engaged him to run demonstrations at his lectures, but by 1803 Accum had left his employ.

Himself rather charismatic and handsome in a bullish sort of way and ten years older than Davy, Accum might well have felt uncomfortable as the young tyro's assistant. Immediately, he published *A System of Theoretical and Practical Chemistry* in two volumes and began to offer practical courses of instruction in his laboratory. He is credited with training many of America's early professional chemists and supplying their universities with laboratory equipment.[19] He produced the first chemistry sets for amateurs, sold in handsome chests. In the next ten years Accum became heavily involved in efforts to establish gaslight in the capital, which began with the lighting of Westminster Bridge in 1813 and included his installation of lights in Rudolf Ackermann's printworks – hence his famous volume on gaslight, which Ackermann published in 1815. But perhaps the greatest contribution to his adopted country was Accum's work on food adulteration, which culminated in 1820 with a famous treatise – attended with disastrous personal consequences.

One of London's less spectacular scientific careers was that of William Martin, the man who would one day denounce Humphry Davy as a murderer. Since John's departure from Newcastle in 1806 William had turned his effervescent mind to a variety of mechanical devices. His plans for a new type of shoe and a lifejacket were both stolen from him.[20] After that he turned his mind to a much more ambitious concept, having been chosen by the Almighty, in a dream, to discover the great secondary cause of all things and the true perpetual motion. Prometheus may have provided the most potent creation myth among contemporary scientists, but William's upbringing and sensibilities led him to search for a more tangible lead in Genesis 2:7: 'And the Lord God formed man of the dust of the ground, and breathed into his nostrils the breath of life; and man became a living soul.'[21] In William's reading the message was clear. If God was the first cause of all things, air was the second. After 36 unsuccessful experiments, he finally engineered a machine which ran unaided. He set it in motion in Newcastle in January 1807 and the following year he exhibited it at 28 Haymarket in London where, among others, the Prince of Wales came to see it. The admission price was a hefty two shillings. William unwisely sold the rights to it in 1810: it was still running 29 years later.

His surviving description of the machine [22] suggests that what William had designed was something like a forerunner of the Stirling engine,[23] using the temperature and pressure differential between indoors and outdoors to turn a tiny reciprocating motor. Sceptics thought that the timepiece which William connected to his machine's pendulum to prove its reliability was making it work, rather than the other way round. He duly removed the timepiece but, even so, many a noble visitor was convinced there was some trickery at work. The Duke of Northumberland even joked with William about it, telling him he would find 'noblemen the greatest fools'. The eminent mathematician Charles Hutton (who had started life as a hewer at Benton Colliery on Tyneside) when he came to see it, kept tapping the railings which protected it, convinced that they harboured secret magnets. When asked if Hutton was right William replied, 'I am sorry to say he is as far wrong as the old woman got when she scratched the bedpost instead of her bottom.'[24]

William was probably in London at the same time as his second brother Richard, stationed there with the Guards until the outbreak of the Peninsular War after May 1808. Jonathan Martin was at this time enduring the horrors of the bombardment of Copenhagen. Both Richard and William appear to have made full use of their youngest brother John's relative domestic stability, sponging off him when they could. But that stability did not last. The glass-painting business was broken up. Charles Muss opened an exhibition in Bond Street, but that failed too and he became a bankrupt. Swallowing their pride, he and John went to work for William Collins, a former rival whose business on the Strand supplied glass to the Royal Family. John was once again paid two pounds per week, but it was hardly a secure existence. He had not served a proper apprenticeship in the trade and despite, or perhaps because of his talent, his presence was resented by some of Collins' employees who had served their time.

Even so, John was confident enough of his own future to marry. His bride was Susan Garrett, a Hampshire woman of whose background little is known. She had been a friend and frequent visitor at the Musses; she might even have been the cause of the 'little differences' that persuaded John to move out of the household. She was older than John probably by five years, perhaps by nine. The Martins' first married home was in quarters behind Collins' shop. From here John eked out his earnings by teaching and selling watercolours. Susan's support for the budding artist, now aged 20, was as much intellectual as emotional. She was highly literate: she read widely and voraciously, often aloud to John so that she became effectively the filter through which he broadened his education. He confined his own reading to Milton and the Bible (and later Byron).[25]

One of John's and Susan's shared passions was for chess: both were expert players. It was probably because of this interest that they became acquainted with the Hunt brothers some time after they launched *The Examiner*, the most famous literary and political magazine of its day. Its proprietor, John Hunt, and his wife Sarah were also devoted to the chessboard. They and the Martins began to meet weekly, alternately at each others' houses, to play matches that were taken very seriously indeed. So much so that mutual acquaintances began to come along to watch. This was the origin of the Martin 'evenings' which later became a celebrated feature of London's social, literary and scientific circles.

The Hunt brothers, John and Leigh, had an odd upbringing. Their father Isaac was a dissolute preacher, former New England lawyer and sometime tutor to the wealthy. His social and financial descent led via the alehouse to the King's Bench debtors' prison. John, the elder of the two boys by eight years, was a politically committed reformist whose early career as a printer developed into that of newspaper proprietor. Leigh, after an enlightening but traumatic education at Christ's Hospital (whose alumni included Charles Lamb and Samuel Taylor Coleridge) became a theatre critic with a reputation for single-minded toughness and impartiality.

The Examiner first appeared in 1808. It was designed as a political weekly, an antidote to innumerable Tory organs such as the *Morning Post* which slavishly supported the war, the king and the establishment. There were other radical journals: William Cobbett's *Political Register* was one of the most widely read papers of the age: splenetic and anti-establishment, but hardly intellectual. In Scotland the *Edinburgh Review* had begun publication in 1802 and was the flagship of the intellectual liberal cause. Advances in printing technology and the use of the Royal Mail as a fast and efficient distribution system fostered an explosion in commercial publishing around the turn of the century. The government, sensitive to its loss of control over information, especially among the labouring classes, responded with punitive stamp duties on newspapers but these did not dissuade their readers, who were able to share copies at coffee houses and subscription libraries.

The wartime government of William Pitt had had its fingers burned during the Treason Trials of the mid-1790s. The so-called Ministry of All the Talents, formed in 1806 after Pitt's and Fox's deaths, was particularly sensitive to any hint of sedition but also wary of the dangers of heavy-handedness. The war dragged on and showed no signs of ending. Public debt was barely sustainable. An ageing, periodically deranged king was being shadowed by a considerably more unsavoury heir in the Prince of Wales. The abolition, in 1807, of the slave trade (though not slavery itself) had not satisfied a reform movement that still clamoured for Catholic emancipation, for political reform and an end to corruption. The hunger for change was growing more for having been held in check so long. In its first three years *The Examiner* attracted three charges of seditious libel.

John Martin's circle of friends eventually included anyone who was anyone in London. It is difficult to know exactly how many of these friendships came about; they are often referred to in reminiscences as already-mature relationships with obscure origins. The Hunts were certainly an early link to a fashionable set of leftish writers, artists and scientists for whom Prometheans seems an appropriate epithet; they had no name for themselves. They were proud of their, for the most part, humble origins. They venerated the Greek Titan's act of insolent theft and its disinterested moral purpose. They were united in their desire for social, scientific and artistic progress. As they gradually drew together in the glow of London's flame these firebringers began to achieve a sort of critical mass.

Charles Muss was probably responsible for introducing the Martins to many of the leading painters who lived in London: Turner, Constable, B. R. Haydon and C. R. Leslie. There were any number of clubs and coffee houses where one might find and cultivate likeminded company, especially in the area around Marylebone where so many of them lived. John was not alone in seeking friends outside his own métier. Davy consorted with poets; Faraday courted musicians and painters; Turner liked the company of scientists and Shelley was a keen student of chemistry and anatomy.

Percy Bysshe Shelley, who was to become the unacknowledged prophet of the cult of Prometheus, had not yet met the Hunts, or probably the Martins. At this time he was not even aware that William Godwin, his future father-in-law, was still alive. Born in Sussex in 1792, he was the son of Sir Timothy Shelley, a baronet. Like Turner he was a prodigy. He displayed a remarkable memory from an early age. He was interested, like his fellow Prometheans, in everything. He was disturbed by poverty and ignorance; his internal universe was filled with chemistry and the classics, ghosts and scientific instruments, with thrillers, with beauty. Schooled at the Syon House Academy and at Eton, his privileged background makes a marked contrast with the Martins, with Turner and Faraday. Whatever else motivated him it was not a wish to escape the gutter. Rebellion, restlessness, paranoia and a love of liberty were his controlling passions. By the time he and his friend Thomas Jefferson Hogg were sent down from Oxford for writing a pamphlet on *The Necessity of Atheism* [26] Shelley had already furnished himself with a complete Promethean toolkit, as Hogg recalled in describing his rooms at University College:

> … books, boots, papers, shoes, philosophical instruments, clothes, pistols, linen, crockery, ammunition, and phials innumerable, with money, stockings, prints, crucibles, bags, and boxes, were scattered on the floor and in every place; as if the young chemist, in order to

analyze the mystery of creation, had endeavored first to reconstruct the primeval chaos ... an electrical machine, an air pump, the galvanic trough, a solar microscope, and large glass jars and receivers, were conspicuous amidst the mass of matter.[27]

Shelley first met and dined with Leigh Hunt in May 1811, shortly after his expulsion from Oxford and before his precipitate marriage to Harriet Westbrook later that year. Almost, it seems, incapable of settling in one place for more than a few weeks or months at the most, Shelley did not meet Hunt again until 1816. He never, in truth, became a comfortable member of the London set that was emerging in the second decade of the nineteenth century.

London, and particularly the fashionable but inexpensive area around Marylebone, had almost the atmosphere of a university town. It was intellectually and socially mobile, partly as a result of febrile wartime edginess and partly because of the liberalizing effects of the Industrial Revolution. For a while, recruitment into an elite group of budding movers and shakers was dependent more on talent and personality than on birth, rank or wealth. Its presence began to discomfit the political establishment.

John Martin's almost touching provincial eclecticism, his child-in-a-sweetshop love of the City, was a distinct advantage as he built a web of social and professional relations careless of sectarian prejudice. His connections and friendships encompassed the radical left, the established Church, conservatives, atheists and Quakers. He was interested in radical politics, in engineering, in literature, sport, fine arts and architecture. He was Byronic in appearance and temperament, passionate in argument, loyal and generous to friends, and he and Susan liked the company of men and women of humour, intelligence and talent. They were a couple worth knowing.

In 1810 John for the first time sent an oil painting, *Clytie*, to the Royal Academy Summer Exhibition. The picture was not hung but the Professor of Painting, Henry Tresham, encouraged John by telling him it had only been excluded for want of room. The Clytie of myth was an ocean nymph, deserted by her beloved Apollo and changed into a heliotrope whose blossom was said to turn its head always towards the sun. In Martin's painting (the surviving version was painted in 1814 and now hangs in the Laing Art Gallery in Newcastle) she appears as a tiny diaphanous figure in the foreground of a blue-hued, wooded plain of great ethereal beauty and charm. She is framed by rocks to either side and lofty mountains loom out of a distant sun-infused haze. The composition and handling of paint gives the whole a dreamy quality. The light is unmistakably Northumbrian.

The diminutive size of the figure and the awesome scale of the landscape laid down immediate and unconventional markers for the rest of Martin's career. He was already bending the rules of traditional historical landscape painting; he soon shattered them. *Landscape Composition*, painted the following year, was hung at the Academy, though in an unfavourable position. It does not survive but was probably a more conventional work – a costume drama as opposed to a cinemascope epic.

At the beginning of 1812 John's budding career as a glass and enamel painter came to an abrupt end when his fellow workers struck against him. He and Susan had just had their second child, Isabella; the first, Fenwick, was two years old. John later told his friend Ralph Thomas that Collins' other painters were jealous of his popularity with buyers and, having discovered that he had not served his full time as apprentice, forced him out. One has to be slightly careful of this third-hand testimony. Sergeant-at-law Ralph Thomas (about whom nothing else is known) was in later life a very close and loyal friend to John. He wrote a diary, apparently ill-organized and ad hoc, which formed the best contemporary record for John's habits, behaviour and thoughts. John's first biographer, Mary Pendered, had access to the diary in the 1920s. She quoted from it at length but when she was questioned about it by Martin's second biographer, Thomas Balston, in the 1940s she replied rather defensively that she could not remember what had become of it. Neither Thomas's family nor Pendered's were ever able to locate the manuscript and the painful suspicion must be that she discarded it.

A passage relating to Martin's resignation from Collins' firm that survives from Thomas's diary via Pendered nevertheless rings true and it must be inferred that she quoted it accurately, even if Martin was remembering what he would have liked to say, rather than his real words at the time. John told Thomas that on leaving the workshop he berated his fellow painters: 'You have done your best to ruin me, and your conduct is tyrannous and unjust. I pity you, as you act in ignorance ... Your blind vengeance will recoil upon yourselves ... My designs have caused emulation and excitement among you that, of itself, has improved your work and gained you better pay.' [28]

Despite his characteristic bullishness, John now faced a bleak future, especially if, like Jean-Paul Marat, he allowed himself to become obsessed by injustice, tyranny and blind vengeance. He was 23, married with two children, and he had no trade to fall back on. He had been ambitious to continue in the glass- and china-painting business. He felt he was achieving new standards in technique and workmanship. Like many fellow-artisans of the period he bitterly resented the strictures of the indenture system. He might have got away with his lack of technical training if he had possessed a more placable temperament. His tone suggests that he was prone to reminding his fellow workers of their shortcomings; so their resentment may well have been more than merely professional.

John might now have chosen to use his growing circle of well-connected friends to set up his own business, although it would have required him to raise capital in a period when this was not such an easy or obvious thing to do. He might, besides, have realized that he was better off working on his own and for himself. He resolved to pursue oil painting as his sole career.

These were defining times for all the Martin brothers. Richard decided to stay with the Guards. He followed Wellington to the Peninsula and through those long years of desperate campaigning, brutality and hardship he survived uninjured – from Talavera to the lines of Torres Vedras, through Busaco and on to Badajoz, Salamanca, Vitoria. None of his experiences are recorded, but enough of his comrades' testimony survives to appreciate the highs and lows of that exceedingly nasty war. By the end of 1812 Wellington's armies were in control of Spain and in the following year, after Vitoria, redcoats set foot decisively on French soil.

As if the Peninsular campaign and Napoleon's Continental System of economic warfare were not enough to cope with, Britain managed in 1812 to provoke the young United States into a pointless and costly naval conflict which, though it lasted less than two years, diverted resources from the European theatre. Its primary cause was mercantile protectionism; its flashpoint the inept searching of American naval vessels for British deserters. It had been allowed to escalate because Britain underestimated the capabilities of America's small but well-armed and highly trained fleet of heavy frigates. In a year of uncertainties at home and Europe-wide conflict, newspaper reports of British ships being sunk and captured were of no help to a government whose Prime Minister, Spencer Perceval, had just been assassinated.[29] In the previous year there had been Luddite riots in Nottingham, the king's mental and physical decline had forced the government to appoint the Prince of Wales as Regent, and Britain was broke. The timing of the war could not have been worse.

George Gordon, Lord Byron, in his maiden speech to the House of Lords in February 1812 during a debate on the Frame Work Bill, summed up the internal strife that threatened to break the country:

> The House is already aware that every outrage short of actual blood shed has been perpetrated, and that the proprietors of the frames obnoxious to the rioters, and all persons supposed to be connected with them, have been liable to insult and violence ... But whilst these outrages must be admitted to exist to an alarming extent, it cannot be denied that they have arisen from circumstances of the most unparalleled distress: the perseverance of these miserable men in their proceedings tends to prove that nothing but absolute want could have driven a large, and once honest and industrious, body of people, into the commission of excesses so hazardous to themselves, their families, and the community.[30]

Jonathan Martin avoided any part in the American war because he had deserted the navy at Cadiz in 1808. His adventures continued, however, as he joined a merchant ship bound for Egypt. Here, corn was bought to resupply Lord Collingwood's Mediterranean fleet based in Sicily, the strategic pivot of the naval war in Europe. Jonathan considered his time in Egypt a turning point in his so-far sorry life. He was overwhelmed by the biblical wonders of the land of Joseph and it led him to reflect how God had preserved him from so many dangers, as if for a purpose. He was troubled by guilt and resolved to amend his life. This did not stop him from getting involved in any number of scrapes which included being attacked by a party of camel drivers when he was left in charge of his ship's boats.

Back in Messina he was nearly killed when some of his mates provoked a drunken fight with a group of Sicilians at a wine-house. Jonathan, by now enjoying born-again sobriety, was able at the last moment to save his messmates by pleading on their behalf that they were drunken English sailors who meant no harm. He was now fully convinced of the regular intervention of a divine hand on his behalf. And on his journey back to England in a vessel carrying Sicilian sulphur for the gunpowder industry he was once again to be grateful. The ship was caught in a terrible storm in the Bay of Biscay in which she broached-to, her rails were torn away and she began to take on water from a sprung timber below the main chains. According to Jonathan's account it looked as though it was all up with them; the ship was on her beam ends and would not answer the helm. But he rallied the crew (apart from an inexplicably cowardly captain), fell to his knees and prayed and between them they managed to right her. In 1810, therefore, after six years at sea Jonathan returned to England. He spent a few days, probably with his brother John, in London, then made his way north to be reunited with his parents. For a while after that his life was less adventurous. He settled in Norton, near Stockton-on-Tees, found employment as a tanner, and married. His only child, named Richard, was born in 1814.

William Martin was no longer in London. After installing his perpetual motion machine he returned to his job at the Howdon ropery. Unlike John he lacked the determination to overcome professional setbacks; but he was still restless, and just as Jonathan was returning to the northeast, William rejoined the Northumberland Militia at Ipswich. He spent some time with a detail guarding French prisoners of war and was then posted to the barracks at Colchester. From about 1811 until 1814, when the militia was finally stood down, William was stationed in County Cork where he found time to learn the art of wood engraving. Some of his efforts, particularly those illustrating Jonathan's adventures, are extremely crude in execution. Others, cartoons of a more or less bawdy nature, reflect the vibrant talent shared by all the brothers.

In 1812 John and Susan Martin and their two children moved from the rear of Collins' shop behind the Strand to 77 High Street, Marylebone. Here they were at least part of a community of aspiring artists and writers. They lived not very far from the Royal Institution on Albemarle Street where budding young men of science were congregating, and within easy reach of open countryside where they could walk and John could think and sketch. He painted his second major oil in one month. Its subject matter leaves little doubt as to the mood he was in: *Sadak in Search of the Waters of Oblivion*, from an obscure Persian myth [31] is outcast man struggling against overwhelming odds. The vertical proportions are more ambitious than those of *Clytie*; the scale is vaster, the mood darker. At 54 by 42 inches the canvas is not large by Martin's later standards; the scale is provided by his trademark diminution of the human figure Sadak, who seems to be climbing literally off the bottom of the frame onto a rocky ledge by a waterfall. The background is constructed of rugged mountains soaked in the livid dark red glow of abysmal fire.

No written document survives to prove that John Martin was a devotee of Prometheus. He may or may not have known about the treatment of the Promethean myth by Fuseli, Blake, Romney and others before the turn of the century. But *Sadak* makes it absolutely clear that he understood and identified with the chained Titan from the beginning, even if he saw the myth in terms of Milton rather than Aeschylus or Hesiod. Milton's Satan, though less equivocal a character than Prometheus, possesses all his creative energy, his potential and his fatalism. [32] John had not perhaps, at this time, yet met Godwin and through him Shelley, though Godwin's influence in diminishing the human form is already apparent. John may in fact never have met Shelley, though it seems unthinkable given their shared connections and the mutual influence that their works display.[33] So while poets and men of science were beginning explicitly to call on Prometheus as their irreligious 'deity', painters like Martin, with his overwhelmingly biblical upbringing, chose to interpret the Satan/Prometheus/Mephistopheles character within a Christian, specifically an Old Testament framework.[34] In *Sadak*, after the perceived injustice of his treatment at Collins', Martin saw himself quite naturally in the role. And perhaps not just himself. It may be no coincidence that *Sadak* was painted in the year that John's friends the Hunts were convicted of seditious libel and imprisoned.

Already, in its second year, *The Examiner* had attracted government prosecutors after publishing an article by Leigh Hunt which exposed the Duke of York's scandalous involvement in the sale of army commissions. The case never came to trial and the emboldened Hunt brothers provoked more censure when they reprinted a libellous article from the *Morning Chronicle*. Again, the charges were dropped. In 1811 a more serious charge did come to court over an article criticizing harsh discipline in the army which appeared to suggest that French soldiers were better treated than their English counterparts. To a wartime government this seemed a clear case of sedition, though in the event the administration had no better luck with the judiciary than it had during the Treason Trials. The Hunts were successfully defended by Henry Brougham, the most gifted, admired and hated lawyer of his age.

Brougham[35] was born in Edinburgh in 1778, the same year as Humphry Davy. From their portraits they might be brothers although Brougham, unlike most of his fellow-Prometheans, belonged to a comfortably landed Westmorland family. Like Shelley, Brougham was a well-bred prodigy, attending Edinburgh University from the age of 14. Unlike Shelley he was a natural academic and thrived at university without being sent down. His interests knew no bounds: history and the classics, science and mathematics, literature and the arts. The range of his publications is staggering; from a student paper on the composition of light to pamphlets against slavery there seemed to be no compass to his talents. It was said by his biographer that if he were locked up in the Tower of London for a year without a book, he would emerge having compiled an encyclopaedia.[36] As a boy of ten, in 1788, Brougham sat on the shores of Loch Dalswinton near Dumfries in Scotland to watch the trials of a radical iron-hulled steam vessel designed by Alexander Nasmyth; one of the passengers was Robert Burns.[37]

In 1802, along with a few friends who included the famous clerical wit Sydney Smith, Brougham founded the *Edinburgh Review* (he contributed 35 articles in its first two years), a standard-bearer of northern liberalism which from 1808 was mirrored by the Hunts' *Examiner* in London. Brougham had been practising law since 1800. A year after founding the *Edinburgh Review* he made the very natural move to London, drawn to its flame like so many others. Here he met Charles Lamb and William Hazlitt, Lord Byron and the Hunts.

In 1810, after a spell with the British Mission in Lisbon, Brougham was offered the parliamentary seat of Camelford by a leading Whig peer, the Duke of Bedford. In Parliament his forensic skills showed him to be a rising star of the liberals but his first term as a member lasted only two years. In the 1812 election he stood for Liverpool and was defeated by George Canning in a famously bitter campaign. Now he concentrated his efforts on defending controversial political cases. A year after his successful defence of the Hunts he acted for 38 Manchester hand-loom weavers who had been arrested for attempting to form a trades union. They too were acquitted.

The Hunts were not so lucky when they were tried a second time. An article in the Tory *Morning Post* had described the recently elevated Prince Regent as 'an Adonis in loveliness', a portrayal which very naturally attracted ridicule since the Prince was notoriously overweight and in his habits and behaviour quite unlovely. *The Examiner* not only satirized his corpulence but accused him of being '… a violator of his word, a libertine over head and ears in debt and disgrace, a despiser of domestic ties, the companion of gamblers and demireps, a man who has just closed half a century without one single claim on the gratitude of his country or the respect of posterity!'[38] This was too much. The Hunts refused to retract or negotiate and were duly tried. In February 1813 they were sentenced to two years' imprisonment: John in Cold Bath Fields Prison and Leigh in Surrey Gaol. Leigh's sorry incarceration was mitigated by the visits of many friends, by public sympathy for him and by generous financial support that allowed him to furnish his rooms with flowered wallpaper and a piano among other little luxuries. There was, perhaps, something quite satisfying in taking the lead role in a very public production of *Prometheus Bound*.

John Hunt's time in prison was much less comfortable: Cold Bath Fields was said to enjoy the severest regime of any prison in England; some called it the English Bastille. Situated in Clerkenwell (where Mount Pleasant post office now stands) it had been constructed to a design by the prison reformer John Howard but by the time John Hunt was placed there it held many more prisoners than planned: some two to three thousand. Inmates had to undertake hard labour on treadmills or pumps – not for any industrial purpose, but simply for their moral correction. An internal view of the prison was included as one of the sights of the metropolis in Rudolf Ackermann's *Microcosm of London*. Strange as it may seem, during the period of the brothers' imprisonment *The Examiner* continued to be produced under Leigh's editorship without hindrance from the government. Its circulation at this time was between seven and eight thousand copies a week.[39]

The severity of *The Examiner*'s attack on the Prince and the government's reaction to it can be partly explained by the effect of the Regency on the reform movement. For more than a decade the Prince of Wales had staunchly supported Charles Fox's opposition Whig party in their campaigns for parliamentary reform, for Catholic emancipation, for an end to corruption in government, and the abolition of slavery. These were the war-cries of *The Examiner* and other radical and reformist organs. Now that the Prince held the reins of power he no longer showed any inclination to support the opposition and the Whigs once again fell into disarray. The Prince's politics had been the opposite of his father's precisely because they were his father's; he himself had no moral political conviction at all. With the Regency, therefore, hope of real reform was smothered for another generation. The reformists naturally enough vented their anger on the man who provided them, in truth, with a generous target. One result of this, with consequences that haunted the Prince on his own accession to the throne, was that the opposition, of whom Henry Brougham was a prominent member, came to raise their standard for the Prince's estranged wife Caroline.

John Martin's first act of moral sedition, *Sadak*, was hung in an anteroom of the Royal Academy where it received notices from the newspapers but was evidently not regarded as a political tract. Critics found its technique lacking and composition flawed. At the end of the exhibition *Sadak* was returned unsold. John, undeterred or perhaps with no other avenue to try, continued to hawk sketches and water-colours with little success while planning his next, more ambitious work in oils. One day, however, not long after the closure of the exhibition, he came home to find that a gentleman had called and left his card with the message that he wished to know John's asking price for *Sadak*. John turned about immediately and tried to find the address on the card but failed. In the morning he tried again and had better luck. The prospective purchaser was William Manning, a Member of Parliament and a governor of the Bank of England. His son, who had just died, had often visited the Academy to admire *Sadak* and Manning wished to have the painting as a reminder of his lost child. He paid Martin 50 guineas, admitting that it was worth more (he was not a banker for nothing) but assuring Martin that he could claim his future support.

John Martin, with his second oil painting, had thus set a pattern for his career: enjoying the appreciation of art lovers who were not critics and the happy luck of finding wealthy patrons to support him. He might now begin to believe that he had a career as a painter ahead of him. But just as his professional prospects began to brighten, dark clouds cast their shadow over the family. In 1813 both the Martins' parents died and John and Susan's four-year-old son Fenwick, named after his grandfather, followed them to the grave.

CHAPTER SEVEN

Peace dividend

DESPITE 20 YEARS OF WAR and the sanctions imposed by Napoleon's Continental System after 1806, intercourse between England and France was never entirely suspended. A packet boat ran regularly between Harwich and the Continent bringing newspapers, letters and offers of peace from one side to the other. *Le Moniteur* and a number of scientific journals were available to enlightened and interested parties in London, just as English newspapers were read at the Tuileries in Paris. Scores of smuggling vessels maintained less official links.

At the Royal Institution Humphry Davy kept a close eye on European developments in chemistry and in particular entered into debate with Joseph Gay-Lussac and Louis Thénard on the composition of acids and alkalis. Davy's experimental brilliance, harnessed to his pioneering use of electrolysis as an analytic tool, had been responsible for a string of discoveries and he had managed to isolate potassium, sodium and other metals in the process of decomposing alkali compounds. In 1807, though it may seem extraordinary to us, he was awarded the Napoleonic Gold Medal for the work on electrolysis and as his fame widened he was much in demand for his expertise. The following year he was asked by the British government to advise on conditions at Newgate Gaol. He visited the prison, saw for himself the ghastly conditions in which inmates were held, advised the governors on a system of ventilation, and caught typhus.[1]

In 1810, having followed a number of false theoretical leads and dogged by experimental glitches, Davy pulled off one of his greatest technical triumphs in isolating chlorine, a substance whose properties had perplexed chemists for a generation. A gas at room temperature, chlorine combines so readily with every other element that it is never found freely in nature. Since its production by Carl Wilhelm Scheele in 1774 it had been supposed that the gas must contain oxygen. Davy's proof that it was an indivisible element cemented his international reputation. Then, as if realizing that his experimental energies were beginning to flag, he began to accept invitations to teach outside the Institution. He gave a series of lectures in Dublin in 1810 and 1811 and became more involved in the management of the Royal Institution. The year 1812 marked the high point of Davy's sensational career in more ways than one. He married Jane Apreece, a much-admired, widowed Scottish bluestocking of substantial means. In April he was knighted by the Prince Regent. He had not only risen far above the station of his birth, a fact which made him a soft target for scurrilous newspaper gossip; he had also succeeded in making science a profession compatible with gentility. With marriage, Davy ended his long and continuous career as a lecturer at the Royal Institution, enjoying at last the social and financial fruits of his long labours.

The last few of Davy's lectures, in the spring of 1812, were attended by Michael Faraday, now in his late teens but still toiling away at Ribeau's bookbinding establishment and absorbing all the knowledge he could lay his hands on. Ribeau was still encouraging his apprentice. One evening he showed some of Faraday's lecture notes from the City Philosophical Society on Dorset Street to a friend whose father, William Dance,[2] offered Faraday tickets to see Davy lecture. Faraday sat in the gallery at the Institution taking copious notes, a habit he had formed to compensate for a terrible memory. He, like so many before him, was entranced and convinced by Davy's dazzling performances.

By the autumn of 1812 Faraday's apprenticeship was nearly at an end. Frustrated, he wrote to Sir Joseph Banks (the botanist who had established his career as a naturalist on James Cook's voyages in the Pacific) asking the Royal Society's president for any position of work connected with scientific research. Banks snubbed him, not bothering to reply. Then, in October 1812, Sir Humphry Davy was drawn back to his laboratory by word from Paris that Pierre Dulong had discovered an explosive compound of azote (nitrogen) and chlorine. The news arrived in the form of a letter from André-Marie Ampère, who warned Davy that Dulong and his investigators had already lost an eye and a finger between them in the cause of science. In the bowels of the Royal Institution Davy repeated Dulong's experiments and obtained nitrogen trichloride but, typically, he took few precautions and was temporarily blinded after an explosion with the new substance left a splinter of glass in his eye.

For a while Davy could neither read nor write. William Dance suggested that Faraday might help him out for a few days, since he wrote with a good clear hand. As a result, the two met for the first time. After this briefest of relationships Faraday sent Davy the bound volume of notes which he had taken at Davy's lectures and asked, begged him for a position. Davy met him again in the early weeks of 1813 but pragmatically advised Faraday that he would be better off in the bookbinding business, so insecure was a career in science. Faraday was desolate.

Now, for a second time, chance came Faraday's way. One of Davy's assistants, William Payne, a sort of chief cook and bottle-washer at the Institution, was sacked after a brawl with the resident instrument-maker. The same evening Faraday received a note from Davy offering him the post of assistant on a guinea a week with two rooms at the top of the Institution, plus fuel and candles.[3]

For a few months Faraday was immersed in the work of the laboratory, proving himself an able and gifted operator. He watched and learned from many distinguished men who came to lecture at the Institute and made detailed notes on every aspect of his and Davy's work. But this blissful existence did not last long. In October 1813 the Davys decided to depart on an extended honeymoon to Europe. They were given passports to travel through the Emperor's domains on the grounds that Davy was a recipient of the Napoleonic Gold medal. When, at the last minute, their valet refused to go with them Davy offered Faraday the position until such time as they could find a new man. Faraday's laboratory career was put on hold for two years; his liberal education was just beginning.

The war against France, now in its twentieth year, brought contradictory forces to bear on British society. Fear of invasion, public debt and industrial unrest created instability and hardship. On the other hand, war's economic imperatives, relatively high levels of employment and the suppression of the reform movement created a climate of economic opportunity and growth. The army and navy did not just absorb surplus labour; they were rabid consumers of produce and of new technology. Nowhere was this dynamic relationship between war and industrialization more pronounced than in the Martins' native northeast.

Since the beginning of the Peninsular War in 1808 the kings of the coal industry had wrestled with the problem of increasing production while at the same time reducing costs. As demand for coal increased, new seams were being won at ever-greater depths and at greater distances from the industry's transport arteries, the rivers Tyne and Wear. At these new depths flooding by groundwater was an ever-present problem and so was the danger of underground explosion from firedamp. Improved power and efficiency of steam pumping engines might deal with the first issue; but engineers and scientists had not as yet seriously addressed the technical difficulties posed by the presence of methane and a lack of adequate ventilation. Although the region's buoyant economy kept employment levels high the ample supply of local labour, especially child labour, meant that human life was a relatively cheap commodity. Year after year, men and boys were brought to the surface crushed, burnt and suffocated in their tens and scores for pit communities to grieve over.

The supply and maintenance of horses, by contrast, was becoming a serious issue. The army's consumption of beasts of burden in Portugal and Spain, together with dramatic rises in feed costs, meant that colliery owners and their managers, the 'viewers', now began to look urgently for a replacement for the horse, whose annual maintenance cost was now estimated at 78 pounds per animal. A single colliery might spend more than nine thousand pounds a year on its beasts.[4] Could not their blacksmiths contrive a steam engine that could pull coal along wagonways?

The idea was seductive: when a horse finished its day's work it required feed, bedding and shelter. An engine only required fuel when it was working; and the fuel, effectively, would cost collieries nothing.

The first solutions were static. Boulton and Watt's rotary engines were adapted to pull wagons up inclined planes by means of winches. Other ingenious men contrived self-acting planes by which the inertia of heavy coal-laden wagons rolling downhill towards riverside staithes might be used to draw empty wagons back up. One of these, the Bowes incline near Gateshead, recently celebrated its 180th working year. The idea of a steam engine that might itself move on rails was not new. Richard Trevithick had built two high-pressure machines that were compact enough to be mounted on wheels and had worked with varying success; but he was not able to exploit their commercial possibilities in the northeast and by the time the coal kings began to look seriously at the locomotive he had more or less given up on them in disgust. The third of his patented locomotives to be built was constructed under licence by the safe-makers Whinfields of Gateshead in 1805 for Christopher Blackett, the owner of Wylam Colliery. In effect, Trevithick had handed over the future of the railways to a franchise. In the event the engine was too heavy and Blackett never took possession of it. The locomotive experiment stalled for another eight years.

In retrospect, a steel-wheeled locomotive running on steel rails looks obvious, or at least can be taken for granted. At the time many engineers saw insurmountable problems. By 1810 most wagonways in the northeast were running along the iron plate rails suggested by William Martin among others. But cast-iron rails were brittle, unable to support very heavy loads, and might not offer enough friction for a locomotive being driven uphill. And in an age when parts are standardized it is easy to forget that during the first years of locomotive engineering, before Henry Maudslay and his successors began making replicable machine tools, every single part of a locomotive was bespoke, more often than not forged in the colliery smithy by men used to making horse harnesses and wagons. Engines were built from beautiful watercolour drawings more like architectural plans than engineers' blueprints. This was a brave new world of empirical science, of trial and error unsupported by the safety net of technical manuals, sets of standard calculations and trusted precedent. In Uniformitarian terms, it lacked a set of analogues.

One solution to the friction problem, pioneered by the Gateshead-born Leeds Colliery viewer John Blenkinsop and engine-wright Matthew Murray in 1812, was to engage a locomotive's wheels directly to toothed rails by means of cogs. Blenkinsop's engine was a partial success but the technology severely limited its potential speed. Another northeast engineer, William Brunton, approached the problem from a more organic direction with his extraordinary *Horse-to-go-by-steam*, a locomotive that pulled itself along by piston-driven mechanical legs that grasped the rails in front of it. It was a brilliant concept. Its consignment to history's technological dustbin resulted not from its inherent flaws as an idea, but because it resulted in the world's first locomotive fatalities, long before William Huskisson was run over by Stephenson's *Rocket* in 1830.

For a year after 1814, when it was built, the *Horse-to-go-by-steam* seems to have worked with some success. But in his enthusiasm to impress at the engine's public unveiling at Newbottle Colliery near Sunderland in 1815 before a large crowd, the driver stoked the boiler to its full capacity and then, fatally, strapped down the safety valve. When the boiler blew up, as it must, it threw shards of red-hot iron hundreds of yards in every direction. Along with the shrapnel the engine's manager was blown a considerable distance and dreadfully mangled. One of the overseers died too, as did John Holmes, a pit-boy who had been given pride of place by the boiler to celebrate his rescue from an underground fire the previous month. Fifty others were killed, wounded or scalded in the blast.[5]

And so it is with the humbler efforts of William Hedley at Wylam Colliery that locomotive histories usually begin. Hedley, the viewer, and his foreman-wright Timothy Hackworth, commissioned three Trevithick-type engines for the owner, Christopher Blackett in 1813–14. Blackett had been interested in locomotives since his Trevithick engine of 1805. Now, with horses becoming ever more expensive to maintain, he became a commercially interested convert.

Puffing Billy, the first of the new engines and the most famous of all early locomotives, is said to have been so named not just because of its distinctive wheeze, but also because Hedley was an asthmatic. *Billy* and his sister engine *Wylam Dilly* were not technically pioneering engines; they were, in truth, not very good. Prototypes, they lacked power and the drive was irregular and jerky. *Puffing Billy* survived, and survives, because Blackett was subsequently reluctant to spend the money on having a better engine built. The historical importance of the Wylam engines lies in their physical survival, their status as some of the earliest working coalfield locomotives and in the markers they laid down for future developments.

Hedley had conducted extensive tests on varying configurations of wheels and rails, and believed that he had solved both weight and adhesion problems. It was a question of compromise. Increasing the number of wheels spread the load on rails and meant that they were less likely to crack; but it also reduced the adhesion between wheel and rail. When other local engineers like William Chapman and George Stephenson began to construct their own versions of these locomotives they were able to build on Hedley's successes and failures. By 1815, when the army's precious horses were riding to Napoleon's nemesis at Waterloo, locomotives were in operation at no less than seven collieries across the northeast.[6] A few men saw them as the future; others were not sure. Whether or not they would survive the end of the war, only time would tell.

These last years of war had seen technical innovation flourishing across Britain. In 1812 a single-cylinder wooden steam-paddle vessel, *Comet*, began operating on the River Clyde in Scotland. In 1814, for the first time, a steam press patented by the German Friedrich König was used to print *The Times* newspaper. The same year Peter Mark Roget, discoverer of the phenomenon of 'persistence of vision'[7] and compiler of the English Thesaurus that bears his name, invented the mathematical slide-rule. This was the year of the first cricket match to be played at Thomas Lord's new ground in St John's Wood, not far from where the Martins were living. And the last frost fairs were held on the Thames and the Tyne after a winter of extraordinary severity (the Hunts, in prison, missed the fun and froze).

Their proximity to the Thames did not save the celebrated sawmills belonging to Marc Brunel and so admired by Sir Richard Phillips when, in August 1814, they were burned to the ground leaving just one wing and the engine shed standing. It was a disaster. The Brunels, living with their young family on Lindsey Row (now Cheyne Walk) in a grand old place later owned by the Martins, were comforted by their many friends, among them the novelist Maria Edgeworth. On receiving that lady's condolences the stoical engineer replied, 'The misfortune is not without its consolation, as I shall now have the opportunity of carrying out many improvements which I had before contemplated.'[8] What became of Brunel's child workers is not recorded.

Brunel's stoicism was to be sorely tried. In 1812 he had built a manufactory for making boots to supply the army's soldiers. He had seen Sir John Moore's Coruña veterans returning to England with their feet bound in rags, and determined to design footwear suitable for a long campaign. He won a government contract to make 400 pairs a day in 9 different sizes. His machinery (once more designed and constructed by Henry Maudslay) carried out 16 separate processes, was operated by invalid veterans, and was widely admired. The Duke of York paid a personal visit to see the manufactory himself. Sir Richard Phillips was impressed too, although he had his moral doubts:

> As each man performs but one step in the process, which implies no knowledge of what is done by those who go before or follow him, so the persons employed are not shoemakers but wounded soldiers, who are able to learn their respective duties in a few hours. No race of workmen being proverbially more industrious than shoemakers, it is altogether unreasonable that so large a portion of valuable members of society should be injured by improvements which have the ultimate effect of benefiting the whole.[9]

Brunel's shoemaking concern might have offset the losses of the block-making mill at Portsmouth and the sawmill after its destruction; but the end of the war also saw the end of the army boot business. Brunel was left in 1815 with a mountain of unsold and unsaleable military footwear. He lost £3,000 on the venture, but was rewarded with a Fellowship of the

Royal Society and his ever-fertile brain began to work on other ventures.

In 1813, somehow managing to obtain leave from the militia in Ireland (perhaps because of the death of his parents) William Martin turned up in London. Walking along Piccadilly one day with his youngest brother John he noticed in a shop-window a weighing machine. Instantly realizing its deficiencies he bought it, took it home (presumably to John's and Susan's lodgings in Marylebone) and designed what we now recognize as a set of spring-balance kitchen scales. Doing away with the weights and balance arm, William got John to draw a machine in which a coiled spring was connected to a needle on a clock-like face that measured the weight of a substance by its depression of a plate attached to the spring. Wary of leaving his plans with anyone who might steal the invention, William tried two clockmakers, of whom he became suspicious, and then tracked down an old watchmaker in Holborn and offered him ten shillings and sixpence if the man would make his machine up there and then. He picked it up the same evening and took it directly to the Society of Arts the next day. In December he received a letter informing him that he had won their silver medal and ten guineas, with which His Grace the Duke of Norfolk would present him the following May. It was the only invention for which William ever received an award. The medal, or a copy of it, was often to be seen in later years, adorning William's greatcoat as he propelled himself through the streets of Tyneside on his own prototype velocipede 'the Northumbrian Eagle Mail', wearing an inverted tortoiseshell, bound in brass with a leather strap, upon his head.

Waterloo marked the end of the first global war. But it was not the first end to the conflict. Wise heads remembered the illusory Peace of Amiens in 1802. Younger heads recalled that in 1814 the war had already been ended once by victory at the Battle of Toulouse. Napoleon had gone into his first exile on Elba, compared unfavourably with Prometheus by Byron for not having had the decency to commit suicide.[10] Long-lost sons had returned home only to be mobilized again after Bonaparte's escape and one hundred days of glory. Now a new world of uncertainty was really dawning. Artists were quick to respond to the prevailing atmosphere. During Napoleon's first exile Turner had portrayed this sense of change in *Crossing the Brook*. Ostensibly a narrative of the transition from girlish youth to womanhood, the underlying message in the painting's backdrop (a wooded vale, an aqueduct, a giant water wheel, a manufactory) is that of rural serenity surrendering to industrial progress.[11] By the time the painting was exhibited at the Royal Academy in 1815, the new world prophesied by Turner was dawning.

The manufacturing, or factory system, established by Richard Arkwright at Cromford in 1771 was much debated. Richard Phillips was not its only critic, Turner not its only chronicler. Successive governments had failed to respond to the political consequences of the new manufacturing towns of the north (Manchester still returned no members to Parliament). It also largely failed to address the needs of their workers. The first Factory Act was not passed until 1819. The only legislation which had so far concerned mill workers was the Health and Morals of Apprentices Act of 1802, which limited textile workers to a 12-hour day and ensured some form of social welfare for Poor Law apprentices. It was not enough. So in the industrial heartlands of the north, just as in London, men of vision were taking matters into their own hands.

In 1813 Robert Owen, a Welsh-born cotton mill manager, published the first part of his *New View of Society*. It offered a Utopian vision of an educated and enlightened workforce, heavily influenced by William Godwin's *An Enquiry* and by the Utilitarian philosophy of Jeremy Bentham.[12] Owen's moral and practical authority was unarguable: he was one of the most successful mill owners of his or any other generation. He was not only dedicated to the welfare of his workers; he was making a lot of money from them.

Born in the year of the great flood, 1771, the sixth child of a saddler and ironmonger, he had become a mill superintendent at 19. His great life's project began when he married the daughter of Scots mill baron David Dale and bought Dale's mill (co-built by Richard Arkwright) at New Lanark. His millennial propensities might be inferred from the date of this acquisition: 1 January 1800. At New Lanark,[13] building on the philanthropic work of his father-in-law, he built the first social housing, inaugurated infant education as we know it and set up a shop to provide his workers with goods at little more than cost price. Visitors from across the world came to marvel at this model enterprise for many decades to come. Owen was the original nineteenth-century paternalist. Venerated by socialists as the founder of the cooperative movement and a driving force behind Chartism, he might also, oddly, be seen as the forerunner of what has been termed 'black-box' management. Just as profit might be made by controlling production in the mills to guarantee predictable output using reliable machinery and constant supplies of raw materials, so might the workforce be honed to mechanical perfection and predictability by early instruction and discipline. The New Lanark enterprise produced biddable workers as efficiently as it produced spun cotton. Owen did not believe, as some Prometheans did, in education for its own sake.

The wealthy, the curious, the opportunistic took advantage of the peace in 1814, as they had after Amiens in 1802, to travel. Among the more or less idle rich who came from across Europe to London in 1814 were a number of princes of royal European blood who, deprived of earlier opportunities, now arrived to pay court to the eligible princesses of the Royal Family. Not the least of them was Leopold of Saxe-Coburg-Saalfeld.

Prince Leopold was born in 1790 at the palace of Ehrenberg in Coburg, the son of Augusta Caroline Sophia and Prince Franz of Saxe-Coburg.[14] The time-honoured strategy of such parents was to marry their daughters as well as they could and train their sons in the roles to which they might aspire: diplomacy, the army, the manners of great courts. To which end Leopold, a quiet, pretty and diligent child, was made a captain in the Izmailovsky Regiment of the Russian army (the regiment that had put Catherine the Great on the throne of Russia and which so distinguished itself under Kutuzov at Borodino in 1812) at the tender age of six. The following year, 1798, Leopold was given a colonelcy and in 1803 was made a major-general at the grand old age of 12. These commands were nominal; after the Austerlitz campaign of autumn 1805 so was Coburg, which had been overrun by Napoleon's army. The Saxe-Coburg lands were appropriated and became part of the Confederation of the Rhine. Life as a whole must have seemed rather illusory for Leopold and his siblings; they could follow no profession except that of dynastic chancers; they had no state and no power. Their royal blood seemed worthless.

In 1807 Leopold and his father went to Paris where, apparently, the desirable young man was seduced by the Empress Josephine's daughter Hortense. There was no future in such a relationship. In the long term the only available strategy was to sit the war out and see who would win. Six years later, in 1813, the time came to choose sides. The Austrian statesman Metternich, in a diplomatic and strategic act of brinkmanship that for once outflanked Napoleon, managed to forge a series of European alliances between Austria, Prussia, Poland and Sweden and bring the French to battle. Even so the alliance failed to crush Napoleon at Dresden in August and only a counterattack against General Vandamme's army at Kulm prevented an ignominious retreat. It was at Kulm that Leopold saw his first serious action when he led a force of more than two thousand horse into action. The fate of Germany hung in the balance; but the Coburgs had chosen the right side. More of the Rhineland Confederation states defected to Metternich and at Leipzig in October, when half a million men took the field in the 'Battle of the Nations' Napoleon suffered a disastrous defeat. Now confronted by powerful enemies on two fronts, with Wellington advancing into France from Spain, Napoleon was only months away from exile on Elba.

In 1814 Prince Leopold led his cuirassiers into Paris as a lieutenant-general, escorting Tsar Alexander and the King of Prussia, to whose court he had become attached. He showed himself an able diplomat in the subsequent treaty negotiations. As part of the victory celebrations and having proved himself a young man of talents, he also joined the Prussian party which visited England that summer. Here he was conspiratorially introduced to Princess Charlotte, only daughter of the Prince Regent and likely future Queen of England. She had previously been rejected by Frederick of Prussia and had refused the suitable but unattractive advances of 'Silly Billy' – William, Prince of Orange. She was something of a problem child: estranged from her mother, neglected by her father and insulated from normal society, she was described by a visitor as 'a mutinous boy in skirts'; spirited and attractive, but slovenly and ill-mannered.[15]

Leopold did not remain in London long enough to press his suit; in any case he was not well placed for such manoeuvres. Such was the press of foreign visitors in London that summer that there was no room for him either with the Prussian contingent or in any hotel. Instead, he lodged in rooms in Marylebone and it was here that he became a friend of the Martins.

There is some disagreement about how this happened. Martin's biographer Thomas Balston believed that Leopold took rooms at 77 High Street in the same building where the Martins lived. Leopold's biographer Joanna Richardson believed that he lived at 21 High Street. Whichever is correct, it is certain that they met and became friends; and when Leopold returned to England in 1816, a much more serious courtier, he became John's most important patron.

At the time of their first meeting John's fortunes were slowly recovering from the loss of his parents and first son. He and Susan now had another boy, named after him. John had tackled his first serious Old Testament theme, *Adam's First Sight of Eve*, and it had been hung in the Great Room at the Royal Academy. Now in the Glasgow Art Gallery collection, *Adam* is essentially a tranquil woodland scene apart from the two tiny figures who meet each other in surprise and delight. John sold it for 70 guineas. Its sequel, the brooding *Expulsion of Adam and Eve*, was hung at the new British Institution but never sold. Its most remarkable feature is the introduction of a Martin trademark: the bolt of lightning which for him represented divine retribution and put an essential seal of sublimity on his paintings. The British Institution, located on Pall Mall, had been founded in 1805 by a group of wealthy patrons to promote British artists, especially those out of favour with the Academy. It excluded portraits and its winter exhibitions quickly became popular with the public, so it was an ideal showcase for Martin's unique and exciting style.

In 1814, the year in which he met Prince Leopold, Martin returned to Ovid for inspiration. He sent *Salmacis and Hermaphroditus* to the British Institution and a new version of *Clytie* to the Royal Academy, where it had previously been rejected. The new version (possibly the one that survives in the Laing Art Gallery in Newcastle) was more mature than his first attempt, both assured and confident. Sir Benjamin West, to whom he showed it, was suitably impressed. Introducing John to another aspiring artist, the American C. R. Leslie, he said, 'You must become acquainted as young artists who will reflect honour on your respective countries.'[16] Leslie who, like Michael Faraday, had started out as a bookseller's apprentice, later became Professor of Painting at the Royal Academy and wrote the first serious biography of the otherwise little known John Constable. In later years John's son Leopold Martin, who as a child sat for him, remembered Leslie as one of his father's oldest friends, 'A tall, gaunt man with dark, hard features, recalling the followers of the founder of Pennsylvania. His dress was at all times black, and never without broad-banded black-and-white socks. He was every bit an American; kindly in manners, good-hearted and friendly, but cold to a degree.'[17]

Clytie was hung in a good position at the Academy's 1814 exhibition. Martin, not being an Associate or Academician, was not allowed the privilege of retouching or altering the painting after it was hung, on one of the five 'varnishing days' set aside for this purpose. J. M. W. Turner was well known to take advantage of these days, when a painting could be seen in its final position, to retouch or even completely repaint his works. In his autobiography Leslie told the story of the day, many years later, when Turner arrived at the Academy to find a painting of John Constable's hung next to his …

> When Constable exhibited his 'Opening of Waterloo Bridge' it was placed in the school of painting – one of the small rooms at Somerset House. A sea-piece by Turner was next to it – a grey picture, beautiful and true, but with no positive colour in any part of it. Constable's 'Waterloo Bridge' seemed as if painted with liquid gold and silver, and Turner came several times into the room while he was heightening with vermillion and lake the decorations

and flags of the city barges. Turner stood behind him, looking from
the 'Waterloo' to his own picture, and at last brought his palette
from the great room where he was touching another picture, and
putting a round daub of red lead, somewhat bigger than a shilling,
on his grey seas, went away without saying a word. The intensity
of the red lead, made more vivid by the coolness of the picture,
caused even the vermillion and lake of Constable to look weak.
I came into the room just as Turner left it. 'He has been here,' said
Constable, 'and fired a gun.' [18]

According to Leslie, Turner returned a day and a half later and changed
the red daub into a buoy.

On the first Monday in May, 1814, Martin was able to visit the
exhibition along with the general public. To his horror *Clytie* had been
ruined. A careless or vindictive Academician had spilled a pot of dark
varnish down the middle of the canvas. Although Benjamin West
offered profuse apologies the damage done to John's stumbling career
was, in his mind, irreparable. He was livid, and from this time his
attitude towards the Academy and its members was permanently hostile.
Although Martin repaired the physical damage and may even have
repainted much of it before re-exhibiting at the British Institution,
the painting never sold; nevertheless it remains one of the most
haunting and luminous of his early works to survive.

The ups and downs continued. In the same year that *Clytie* was
ruined the Martins lost their second son. A third son, Alfred was born.
Again, John paused to consider which direction his career ought to
take. Again, he came to the conclusion that he must follow his instinct
and attempt another grand historical canvas. He appears to have been
surviving financially by selling sepia drawings and undertaking regular
teaching work. Probably as a result of his brief friendship with Prince
Leopold he had been recommended to Princess Charlotte's father,
the Prince Regent, as her drawing master. This may be inferred from his
acquaintance with her one-time tutor, the Bishop of Salisbury, who
tried in vain to dissuade John from tackling the next of his subjects,
the hugely ambitious *Joshua Commanding the Sun to Stand Still upon*

Gibeon. A fully worked-up oil sketch of the subject survives in the Ash-molean Museum at Oxford; smaller and less detailed than the final version (which for many years hung in the United Grand Lodge of Freemasons in London before being granted an export licence) it nev-ertheless conveys the sublime mood and ambitious lighting with which Martin intended to make his mark.[19]

It may not have had a terribly racy title, but *Joshua* was the making of John Martin. At 88 by 58 inches it was a suitably large canvas for the scale of composition in which architectural magnificence was for the first time the overwhelming feature. John had let loose his imagina-tion on the biblical city. Rising from a vast plateau behind lowering crags, *Gibeon* (in real life the little village of El Jib, north of Jerusalem) showed that Martin could outdo Bruegel and Altdorfer in portraying the palatial magnificence of biblical decadence. Combined with his diminished human figures in the foreground, a terrible sky and deus ex machina sun illuminating the whole in eerie shafts of light, this was the very essence of sublimity.

Bruegel has surprisingly never been cited as an influence on John Martin. Both had the same approach to human figures: often dimin-ished, concerned with form rather than character and frequently used to evoke not just biblical scale but also the theme of hubris punished. Bruegel's *Landscape with the Fall of Icarus* is nothing if not Promethean. In his architecture Martin echoes *The Tower of Babel* (1563). *Joshua* and *The Bard*, Martin's next great canvas, are reminiscent of *The Conversion of St Paul* (1567) and both painters were keen to portray the battle between good and evil.

Reflecting Martin's declining relations with the Academy, *Joshua* was only hung in the ante-room at Somerset House in 1816. Critics, including Charles Lamb, thought it unrealistic, the miracle of Joshua's command a mere 'anecdote' of the day.[20] *Joshua* also broke with con-vention by elevating the landscape above the subject's human interest and by allowing the weather to play an active, dynamic role in the nar-rative. A twenty-first-century audience brought up on cinematic inter-pretations of cataclysmic events and with an appetite for ever more impressive visual effects, takes such dynamism for granted; demands it.

John Martin was working in cinemascope a hundred years ahead of his time, and the picture did not sell. Nevertheless Martin's public, thirsting for sensation, ensured that *Joshua* was a popular triumph. The subject matter struck a chord, as Martin believed it would, with a public tired of war but excited by wartime accounts of travellers to the Middle East. Apart from published accounts of the magnificence of the ruins of Egypt and Babylon, John would have heard first-hand accounts from his brother Jonathan; his imagination did the rest. The following year, when *Joshua* was re-exhibited at the British Institution, it won the hundred pound premium. More importantly, perhaps, its popular appeal gave the artist heart: 'The confidence that I had in my powers was justified, for the success of my *Joshua* opened a new era to me', John later wrote.[21]

At almost exactly the same time that *Joshua* was hung at the Academy in May 1816 John's friend Prince Leopold married his drawing pupil Princess Charlotte. As the only daughter of the Regent and his consort Caroline, Charlotte, then 20 years of age, was heir apparent to the English throne. While staying in Berlin earlier that year Prince Leopold had been summoned by the future George IV and offered her hand. He had not been George's first choice but he was politically a safe bet and, fortunately for the Court, Charlotte did not object as she had to earlier suitors. Whatever dynastic interests played their part in this arrangement (and Leopold in that sense was no great catch) the simple truth seems to have been that Charlotte was in love with Leopold, as he had been with her since their meeting two years previously.

A princess marrying for love was a rare event, enthusiastically celebrated in London. Leopold was decorated with the Orders of the Bath and Garter and made a general in the British army. Celebrating Leopold's great luck, his chief advisor, Baron Stockmar, was said to have remarked at the time, 'My master is the best of all husbands in all the five quarters of the Globe; and his wife bears him an amount of love, the greatness of which can only be compared to the English national debt.'[22] The marriage made for good, uplifting news in a year of scandals. The Regent's moral and financial excesses had sickened everyone, including the fawning sycophants with whom he surrounded himself. Lord Byron's very public marriage breakdown scandalized London. Shelley's pregnant first wife Harriet Westbrook drowned herself in the Serpentine River in Hyde Park even as the poet was living with William Godwin's daughter Mary. London needed good news, and the royal couple provided it.

Charlotte and Leopold went to live in Camelford House on Oxford Street, opposite Park Lane, with another house in Surrey and an allowance of sixty thousand pounds a year to get by on. Charlotte, the daughter of a disastrous marriage and with no suitable family models to follow, was what Victorians called wilful: ill-behaved, rude, shrieking and immature, though intelligent and attractive. Leopold was cold and formal, his diplomatic training tried by her lack of grace and deportment. Somehow they were perfectly matched and became a model couple.

For a government struggling to cope with postwar unemployment, Luddism, rioting in London and the Regent's embarrassing lifestyle, the Saxe-Coburgs were a much-needed tonic. Soon after their marriage they appointed John Martin as their official historical painter and in 1817 when John's fourth son was born he was christened Leopold after his godfather. As a professional artist, at the age of 28 and after 11 years in the capital, John had made it to the big time. On the deaths of George III and the very unhealthy Regent he would occupy one of the most important positions in the art establishment. He would, no doubt, do Their Majesties sublime justice.

It was not to be. In the first months of marriage Charlotte miscarried twice. Early in 1817 she became pregnant again. Her diet was interfered with by doctors who worried that her fragile body might not carry a baby to full term. They bled her and half-starved her. On 4 November 1817 a child was stillborn. Two days later Princess Charlotte died from a haemorrhage exacerbated by medical incompetence.

Leopold, the Prince Regent, the entire Royal Family and the population of England indulged in an extraordinary outpouring of grief for this childlike future queen, her family's apparent saviour. The death of George III's only legitimate grandchild precipitated a constitutional crisis; it left Leopold bereft and his newly appointed historical painter high and dry.

The constitutional position of the Royal Family now looked bleak. George did not seem likely to have another child and now considered divorcing Caroline and remarrying, a matter looked upon with horror by most of the political establishment. On what grounds would this adulterous libertine gain a divorce? And if he did, his reputation was sure to be dragged kicking and screaming through the courts and newspapers by Caroline's lawyers. His next oldest brother Frederick, the grand old Duke of York, had no legitimate children; nor, as yet, did William, Duke of Clarence. Edward, the Duke of Kent, was unmarried.

The government was at its most vulnerable since the mid-1790s. More than three hundred thousand servicemen had been demobilized after Waterloo and unleashed on an economy precipitated into sudden stagnation. Brunel's shoe manufactory was only one of hundreds whose goods were no longer required as the government sought to halve its expenditure. Food, clothing and armament contracts were rescinded. Prices of raw materials and goods fell almost to levels before the start of the war in 1793. Capital loans could not be repaid. Wages followed prices. The deflationary spiral struck right across the economy. In towns and cities workers rioted because their jobs had been lost. They blamed Irish immigrant workers and the government; they vented their fears and anger on machines. In rural areas low-grade land which had been brought into cultivation during the war now lay fallow and grain prices dropped from more than six pounds a quarter to a little over three pounds.

In wishing to protect its parliamentary interest, that is the landowners, Lord Liverpool's Government passed a Corn Law which guaranteed high grain prices by banning imports until the domestic price had reached four pounds a quarter.[23] For a hungry population the government's effective shoring up of bread prices seemed like heartless provocation and any number of protest groups made ready to march under the white hat of radicalism or the tricolour of the Revolution.

Landed interests also seemed to be served at the expense of the impoverished by the abolition of Income Tax, introduced by Pitt to finance the war debt. The tax, hated as it was, was now replaced by a range of duties on domestic items like salt, beer and candles which only a government hopelessly out of touch with the country's mood could have imposed so witlessly.

There was a genuine revolutionary response across Britain to the government's seeming callousness, but the number, competence and leadership of these groups were limited. In 1816 and the years following, the government used its extensive network of spies and agents provocateurs to infiltrate and sometimes foment their networks. In 1816 there was a serious riot at Spa Fields in Clerkenwell after a meeting to raise a petition was addressed by the radical orator Henry Hunt (no relation to the *Examiner* Hunts). It came to nothing. In 1817 the so-called March of the Blanketeers from Manchester, protesting about the state of the cotton industry, petered out before it reached the capital. The same year an 'insurrection' in Derbyshire led to three hangings and eleven transportations among its supposed leaders. There was widespread (and justified) belief that a government agent named Oliver had in fact been behind it. Across the country radical clubs met to share newspapers and pamphlets, raise petitions and call for parliamentary reform, for more jobs and an end to high prices. In January 1817, even before Charlotte's death plunged the Crown into crisis, the Prince Regent, Thomas Creevey's 'Prinny', was attacked: his carriage was stoned (or perhaps shot at) on its way to the opening of Parliament. The same morning disaffected naval officer Lord Cochrane took a petition of half a million signatures to the House to demand reform and universal male suffrage.

The government's public response in 1817 was to pass three laws which came to be known as the Gag Acts: habeas corpus was once more suspended; Pitt's Seditious Meetings Act was revived; and the Home Secretary, Lord Sidmouth (the former Prime Minister Henry Addington), issued orders that all printers, writers and demagogues responsible for seditious or blasphemous material be arrested.

Its proprietors having been released from prison in the spring of 1815, and convinced that 'some sort of a conspiracy existed ... some actual revolutionary design',[24] *The Examiner* did not hesitate to go on the attack. It compared the fate of 'poor wretches huddling together in ragged starvation on a bridge at night' with the 'high-living prince who has his coach-window cracked'.[25] In March 1817 the Hunts continued: 'There is not a youth … that ought not to blush at seeing a nation, renowned for every species of literature and gentleness, governed against its will by a junto who neither feel what is English, nor can even talk it. – Reform, Reform, Reform: – Petition, Petition.'[26] Percy Bysshe Shelley, soon to become an intimate friend of Leigh Hunt, wrote a pamphlet proposing a means of putting reform to the vote. Hunt and Shelley both wrote pieces on the death of Princess Charlotte, expressing their personal sympathy while at the same time deploring the fate of thousands of her poor, starving countrymen. The atmosphere of righteous anger, with no realistic political leadership to call on, found expression in a form of nationalism. There was talk, as there had been in the reign of Charles I, of Magna Carta. King Alfred's noble fight against the Danes and his princely humility in culinary matters struck wistful chords, as did the Green Man, King Arthur and Robin Hood. Not for nothing did Walter Scott's *Ivanhoe* appear within two years.

John Martin found himself inspired by Thomas Gray's poem 'The Bard' (1754) which celebrated the supposed martyrdom of Welsh harpers by Edward I's invading army. Larger even than *Joshua* and striking because of its portrait format, *The Bard* shows a lonely, isolated figure, a white-maned druid, perched atop a towering crag:

> On a rock, whose haughty brow
> Frowns o'er old Conway's foaming flood
> Robed in the sable garb of woe,
> With haggard eyes the poet stood;
> (Loose his beard, and hoary hair
> Streamed, like a meteor, to the troubled air)
> And with a master's hand, and prophet's fire,
> Struck the deep sorrows of his lyre.[27]

An eagle wheels above him in Gray's own conscious reference to the Promethean myth. In a gorge (a moral chasm?) below, Edward's army marches towards a mountain fastness modelled on Harlech

Castle in an act of crushing military domination. The Bard will shortly leap to his martyrdom.

The Bard might be Martin; King Edward might be the Royal Academy. Equally, Harlech might represent free speech, the bastion of liberty defended by lonely voices in the press (John's friends the Hunts) against the judicial armies of Lord Ellenborough and his Special Juries.[28] This political interpretation (the poem itself begins with the line: 'Ruin seize thee, ruthless King!') is supported by another work of 1817, a pen and ink drawing of *Diogenes and Alexander*, now in the Ashmolean Museum. It catches the moment when the conqueror visits the hermit in his tub. Asked if the mighty Alexander might do the humble philosopher a favour, Diogenes pleads with him to get out of the light. Martin identified himself with the ascetic and cynical philosopher while his figures of authority (despite royal patronage) were tyrants.

From the pen portraits in his son Leopold's anecdotes and from the testimony of his more conservative friends, John Martin might have been a pure romantic. But it does not take a great leap of faith to see him as a rather subtle, politically aware satirist.

The Bard was hung in 1817 at the Academy, high up on a wall in the Great Room, remaining unnoticed and unsold (it now belongs in the Laing Art Gallery, Newcastle upon Tyne). But John, like others tapping into a need for a comfortable vision of England, did manage to make some money that year: first, when Rudolf Ackermann published seven of his etchings illustrating native trees with characteristically tiny human figures peopling the landscape below them. Second, John had travelled little since his arrival in London, but this year he took up a commission he had been offered and visited the Cotswolds where Sir Charles Cockerell had built a capacious Hindu-influenced country house, very much in keeping with the style of the new Royal Pavilion at Brighton and with gardens designed by Humphrey Repton: Sezincote House. Martin's skill in portraying landscape and architecture in novel ways attracted Cockerell to him and he was asked to make some etchings and aquatints of the house and its grounds. A series of sketches, etchings and prints survive from this lucrative excursion from which John gained more confidence ahead of his next great undertaking.

The only one of John Martin's sketchbooks which survives includes many examples of his work at Sezincote and dates from 1818, the year when he, Susan and their children moved to a house on Allsop Terrace on the New Road, barely a quarter of a mile from their old lodgings and situated on the southern edge of what later became The Regent's Park. Some pages of the sketchbook bear studies of carriages with a novel steering mechanism, ship's rudders, what appears to be a breach-loading naval gun and various other mechanical devices which show that John's mind was never solely occupied by fine art. There are designs for pillars, staircases and decorative friezes meant, not to be built but to be incorporated into future pictures. For him architecture was an aspect of engineering and both were the servants of art; later catalogues of his paintings always included scales and measurements of buildings to demonstrate that they were plausible structures. He later admitted that he would rather have been an engineer than an artist and the evidence he gave to various Parliamentary Commissions on engineering and scientific matters show that his interest was not idle.

John Martin carried that interest almost to the extreme in his schemes to bring water and railways to the capital in the 1830s and 1840s. Nevertheless, it is not grand engineering schemes, nor the Sezincote sketches, which make this little book such an important document. On one of its pages is an image whose significance stands out in sharp relief from all the others. It is, or appears to be, a design for a miner's safety lamp constructed in 1818 by his brother William.

The safety lamp is the supreme Promethean icon of the Industrial Revolution, more evocative even than electricity or the steam loco-motive; for its invention (its supposed invention) by Humphry Davy stands as a disinterested act of the highest order, giving light and life to generations of men and boys toiling underground in conditions of neglect, filth, danger and exploitation. John's and William Martin's roles in the history of the lamp reveal a more complex truth.

CHAPTER EIGHT

A light in the darkness

IN MAY 1812, as *Sadak* was being exhibited at the Royal Academy in a London still buzzing with the novelty of the Regency, a devastating explosion tore through the pit at Brandling Main Colliery, Felling, on the south bank of the River Tyne in the parish of Jarrow. The Reverend John Hodgson, vicar of the Venerable Bede's ancient monastery church of St Paul's, described the day's events:

> A slight trembling, as from an earthquake, was felt for about half a mile around the workings; and the noise of the explosion, though dull, was heard to three or four miles distance, and much resembled an unsteady fire of infantry. Immense quantities of dust and small coal accompanied these blasts, and rose high into the air, in the form of an inverted cone. The heaviest part of the ejected matter, such as corves, pieces of wood, and small coal, fell near the pits; but the dust, borne away by a strong west wind, fell in a continued shower from the pit to the distance of a mile and a half ... In the village of Heworth, it caused a darkness like that of early twilight ...[1]

One hundred and twenty-two men and boys had been in the mine when a pocket of firedamp (methane gas) was detonated by a candle or spark. The force of the blast blew the frames from the shafts of both pits, 'William' and 'John'. The shaft lining caught fire. The coal dust ejected from William Pit lay three inches deep but soon burnt itself to cinders while baulks of red-hot timber, blown up the shaft, rained down around the pit-head in a lethal volcanic shower.

Underground explosions were a fact of life in the coalfields. In the previous few years ten men had been killed in a blast at Killingworth Colliery across the river; 4 men and 21 horses had died at Harraton and another explosion at Killingworth had killed 12 more. Colliery communities knew only too well what to expect when they heard the dull thud of a blast and felt the ground tremble underfoot; but the Felling disaster of 25 May 1812 was destructive beyond comprehension:

> As soon as the explosion was heard, the wives and children of the workmen ran to the working-pit. Wildness and terror were pictured in every countenance. The crowd from all sides soon collected to the number of several hundreds, some crying out for a husband, others for a parent or a son, and all deeply affected with an admixture of horror, anxiety, and grief.[2]

The scale of the disaster was magnified because the explosion occurred in the short space of time when both shifts were underground together as they changed over. Only 32 men and boys were brought out alive, of whom 3 died later of horrible injuries. The rest of the bodies lay interred in the workings for more than six weeks before the shaft could be reopened and the corpses removed for interment.

The Reverend Hodgson's account of the Felling catastrophe was controversial. It was the custom in the Northern coalfields for newspapers to avoid mention of colliery disasters for fear of causing offence; and it was in the interests of the colliery owners that they should do so. Against the wishes of the proprietors at Felling, Hodgson proceeded to write a number of reports for the *Newcastle Courant* and in January 1813 he published a full account of the disaster. Part of it was inserted in that year's *Annals of Philosophy* published by Sir Richard Phillips.[3]

London's million fires consumed Newcastle coal at a rate of several million tons a year. Most of her citizens had only a vague notion of the grim reality of mining and the Felling disaster may have been the first great loss of life to have been reported widely in the capital. Even so, with news of the war on the Iberian Peninsula approaching its catharsis, Napoleon's epic march on and retreat from Moscow, the Hunts' libel trial and the murder of the Prime Minister, Spencer Perceval, its echoes rang only faintly.

However, a compassionate barrister named J. J. Wilkinson (who may originally have come from the northeast) was stirred to action. During the long legal vacation of summer 1813 he visited the northeast to see conditions for himself and determined to raise public awareness of safety issues in the mines. On September 1st that year he published proposals for the establishment of a Society for the Prevention of Accidents in Coal Mines. The first meeting of the Society was held in Sunderland on October 1st under the presidency of Sir Ralph Milbanke. Like William Wilberforce's campaign to end the slave trade, it is an early sign of the disinterested humanity which became such a feature of the mid-nineteenth century, quite different in character from the pious Georgian mercy which pardoned condemned criminals or gave alms to the poor.

The Society issued its first report in November 1813. The proceedings contained a letter from John Buddle, viewer of Wallsend Colliery and by wide repute the foremost mining engineer in the coalfield, in which he gave the committee the benefit of his expertise on mine ventilation and urged that efforts be concentrated on 'some method being discovered of producing such a chemical change upon carburetted hydrogen gas as to render it innoxious as fast as it is discharged'.[4]

Buddle's carburetted hydrogen was the inflammable gas known to miners as firedamp and to modern chemists as methane. Firedamp was one of a triumvirate of mining evils. Chokedamp, or carbon dioxide, was the odourless and deadly gas which built up on the floors of coal seams and suffocated miners so quickly that they could not even raise the alarm. Firedamp snuffed out naked flames when concentrated and exploded violently when mixed in the right proportion with oxygen, or 'atmospheric air'. The third evil was water. All three presented safety and engineering problems which increased proportionally to the depth of mine workings in the late eighteenth and early nineteenth centuries.

Coal mines, from Scotland to Shropshire and the Welsh valleys, had to deal with the ingress of water as soon as they penetrated below groundwater levels. Early mines used adits, or horizontal tunnels, to drain mines into natural streams and ventilate them. When shafts became too deep for natural drainage they were lined with waterproof wooden sheathing (later with iron tubing or 'tubbing') and various types of pumps were employed. At first these were bucket-and-chain arrangements of varying sophistication, powered by horse-gins or water wheels. Since the beginning of the eighteenth century Thomas Newcomen's steam-driven pumps, originally designed for the Cornish tin industry, had been in widespread use. At Walker Colliery on the Tyne in the 1750s William Brown built a massive Newcomen-type engine whose cylinder was 74 inches in diameter with a stroke of 10½ feet. By 1813 the much-improved engines of Boulton and Watt were being used both for pumping water and for raising coals. In the rapidly expanding coalfield of the Tyne basin this meant pumping and hauling from depths of over a hundred fathoms, or six hundred feet, but the technology of drainage and engines was by then comparatively well understood.

One colliery engineer, George Stephenson, was earning himself a reputation as an engine 'doctor', able to fix, repair and improve engines to keep them running longer and more cheaply. Only now, with the stakes raised by Trevithick's high-pressure steam, was it becoming important for young colliery engineers to receive some form of technical training; most, like Stephenson, had learned their trade in blacksmith's forges at their fathers' sides. The new locomotives appearing at Wylam proved much more demanding of skilled attention than the static beam engines that drew water and coal. Stephenson himself had been building locomotives for 15 years before he and his son produced the famous *Rocket* in 1829.

In the meantime mine ventilation presented engineers with ever-increasing problems. Firedamp and chokedamp were invisible, chokedamp odourless. Pockets of these gases might build up gradually to deadly levels, or the working of a seam might expose a 'blower' at any time. As shafts were sunk deeper and levels worked

117

more extensively atmospheric air came in shorter supply. In response, any number of enterprising men (William Martin among them) designed schemes for ventilating mines and counteracting the dangers of gas accumulation. One widely adopted solution (with what seems to the modern mind an obvious drawback) was the fire-lamp, a furnace built at the base of the pit's main shaft to draw air through the workings via distant ventilation shafts. The increased airflow resulting from fire-lamps, controlled in various parts of the workings by a complicated series of air-tight doors 'manned' by small boys, was an effective means of diluting gas concentrations. But the more extensive the workings, the more risk there was of the system failing: small boys might fall asleep on the job; doors might jam; and the gradual slumping of overlying strata might render the whole system useless. At Walker Colliery, where the main shaft and the ventilation shaft were eight hundred yards apart, fresh air had to travel more than thirty miles through a labyrinth of seams and tunnels to pass from one to the other.

Quite apart from the evident risk of furnaces themselves causing explosions was the perpetual problem of lighting a mine. Miners had always used candles. They might be used safely in a mine for years before an explosion was set off without warning, though frequently they were extinguished by gas so that hewers and pullers were left in total darkness, not daring to relight them. At collieries where firedamp was a known problem 'firemen' were employed. Dressed in sacking soaked in water they entered the workings in the morning when the mine was empty, deliberately set light to any gas present and, lying flat on their bellies from long practice, hoped to avoid too serious a scorching when a methane flame roared over their heads and up the main shaft.

In about 1740 a mining engineer named Carlisle Spedding came up with a practical lighting solution which could not be extinguished by gas. It was introduced into the mines of his employers, the Lowthers, at their Whitehaven pits on the Cumbrian coast. This device was the steel mill, essentially a grinding stone rotated by a hand-crank against a steel. This gave off a shower of sparks which, though pathetically inadequate in terms of its candle-power, just about enabled pitmen to see what they were at. It was thought, after having been in use for several years, to be proof against the ignition of firedamp. Spedding was using one himself when it lit a pocket of firedamp in 1755, killing him.

John Buddle, the 'viewer' or manager at Wallsend, was only too familiar with every problem that mining could throw at an engineer. His father, John Buddle senior, had been appointed as viewer there in 1792, inheriting a mine whose coal was widely regarded as of the finest quality but whose sinking had been attended with the most awful difficulties. The first shaft was sunk there in 1778, the year of Humphry Davy's birth. It was immediately lost in quicksand. A second shaft was sunk and began to produce coal in 1781 from the Main seam at a hundred fathoms' depth. Then in October 1783, a blower having ignited and set light to the seam, the level had to be flooded to extinguish the fire. Two years later another blast in the same pit forced the viewer to flood the level again. This time an exploratory party, sent down to assess the damage, was killed when their steel mill ignited a pocket of firedamp. There were further blasts in November and December the same year. Subsequently, an arrangement of mirrors was tried, to reflect sunlight down the 600-foot shaft and along the level; but without success. Then phosphorescent fish were brought down, their dim glow employed to little effect.

There were further explosions in 1786 at both A and B pits, all caused by sparks from a steel mill. The third of these killed six men. It was only the sinking of several new ventilation shafts and the construction of more fire-lamps that solved Wallsend's problems for many years ahead.

In 1806 John Buddle junior, aged 33, succeeded his father as viewer at Wallsend. Later he became a consulting engineer to many other local collieries and not for nothing did he earn the sobriquet King Coal. In that year, even as John Martin was heading south towards London (probably as a passenger in a collier) a series of explosions took a heavy toll of life in the Tyne and Wear coalfield. Buddle began to concern himself with improving ventilation. He was especially concerned with fire-lamps, the furnaces sited at the base of shafts to draw fresh air into the workings. The most obvious drawback of this method was that concentrations of flammable gas would be drawn over the furnace, flashing into lethal sheets of flame. When these fires roared up the 600 feet of the main shaft they might set fire to the timber lining; frequently they blasted the superstructure at the surface into the air, along with any miners who might be in the way. At this time miners were still being drawn precariously up and down shafts by clinging to the ropes which hauled corves, or baskets of coal, and waste.

When a fire occurred the only recourse was to shut down the ventilation system entirely and extinguish the furnace by diverting water down the main shaft. Apart from the costly suspension to hewing operations, relighting the furnace was no easy matter, as Buddle himself found out when he spent six weeks at the base of the shaft at Hebburn Colliery trying in vain to relight the furnace with tar and a red-hot iron ring.[5]

Buddle made experiments with steam ventilation and an air pump, neither of which could draw the required volume of air down the pit, before implementing a panel-work system (designed by Thomas Barnes for the nearby Walker Colliery) which became the standard method. In panel-working, areas of the workings were isolated from each other with walls of unhewn coal and ventilated by divided currents, using a 'dumb-drift' shaft so that air drawn up to the surface did not pass over the furnace.

Even with new ventilation systems spreading through the coalfield, finding a solution to the lighting problem exercised the minds of mining engineers everywhere. First to propose the construction of a light in which a flame was isolated from the surrounding air was Dr William Clanny, a Sunderland physician. In 1811 he constructed a 'blast' lamp of strong glass containing a candle. The lamp was enclosed at its base apart from a small tube through which air could enter, forced by a pair of bellows. Its drawbacks were soon obvious. In the rough conditions below ground the glass broke and it required a boy to be employed constantly at the bellows. By the beginning of 1813, in response to the Felling disaster, Clanny had built a stronger lamp whose design incorporated a water reservoir through which air was drawn, to further isolate the flame from potential gas concentrations. On May 20th of that year his paper 'On a Steady Light in Coal Mines' was read to the Royal Society in Somerset House, London. Clanny became an enthusiastic member of the new Society for the Prevention of Accidents in Coal Mines.

One of the Society's leading lights was the Reverend Dr Gray, then Rector of Bishopwearmouth. Possibly at his suggestion, because he had met the great scientist, and in the light of Buddle's plea for a chemical solution to the firedamp problem, the Society's members decided that a formal approach to Sir Humphry Davy should be made to enlist his help. J. J. Wilkinson, the founder of the Society, duly called at the Royal Institution on Albemarle Street, only to find that Davy was absent. He was, in fact, in Paris on his honeymoon, accompanied by his wife and their now-reluctant valet Michael Faraday. Wilkinson wrote to Davy there but because he had neglected to pay the postage the letter was returned and Davy knew nothing of it for nearly two years.

Davy and his assistant had very different experiences of their time on the Continent. Davy's safe passage was guaranteed by the Emperor; he was held in high esteem by French scientists and introduced into the highest levels of Parisian society. Faraday was nothing more than a servant in the eyes of all except Davy himself. In October 1813, as the new Society was meeting for the first time in Sunderland, the Davys and Faraday were touring the galleries of the Louvre, filled with classical and Renaissance wonders largely looted by Napoleon's armies from the courts and private collections of his conquered territories. Davy, apparently, was insensible to the art, remarking only on the fine frames, though his later interest in the chemistry of paint may have been kindled here. Faraday, a much more sensitive observer, was both overwhelmed and disgusted by what he regarded as booty stolen from the capitals of Europe by a despot.

On November 2nd, in the presence of Georges Cuvier and André-Marie Ampère, Davy was presented with the Napoleonic Gold Medal at the Institut de France – even as the Duke of Wellington was forcing a passage into southern France across the Pyrenees. Davy then plunged into the business of the scientific community, most importantly the identification of an exciting new element known only as Substance X. By the time he left Paris the substance had been isolated and he had named it iodine. It was another triumph, although a report in *The Examiner* made some very unflattering remarks about this unpatriotic visit: 'He may talk about so many chemical intentions as he pleases, but he goes to see and to be seen ... to have it said, as he moves along through smiles of admiration and shrugs of obeisance,—"Ah, there is the *grand philosophe, Davie!*"'[6]

Michael Faraday occupied his time by writing a meticulous journal and in seeing as much of Paris as he could. One day he saw the Emperor (a man similarly diminutive at five feet four inches) pass in his carriage. On another he took a considerable risk by taking notes on the operation of the military semaphore on Montmartre. He avoided Lady Davy as much as possible. She was patrician and despotic and treated him like a skivvy. Sir Humphry played a more pedagogic role, involving Faraday in his thoughts as they developed and discussing the geology of the countries through which they passed. In Italy they met Alessandro Volta and visited Vesuvius, which had been more or less active for the last 20 or 30 years. In Florence they saw great lenses being used in an attempt to burn diamonds to pure carbon.

While in southern France, Davy wrote what now seems a prophetic poem which shows not only that he had his own views on the war, but also that the Promethean flame still burned within:

> Albion, thee I hail!
> Mother of heroes! Mighty in thy strength!
> Deliverer! From thee the fire proceeds
> Withering the tyrant; not a fire alone
> Of war destructive, but a living light
> Of honour, glory, and security, –
> A light of science, liberty and peace![7]

Davy did not return to his position at the Royal Institution when his European tour came to an end in the summer of 1815. He believed, perhaps, that he was sufficiently eminent that he had nothing left to prove. His health had suffered in the cause of science, from explosions and the inhalation of gases, from overwork and stress. Experimentally he had achieved great things and his social life now assumed much greater importance. Having single-handedly elevated the natural philosopher to a new aristocratic status all its own he may have wished to become, in effect, Prometheus emeritus.

Michael Faraday became assistant to Davy's replacement William Brande at the Royal Institution, taking up rooms again in Albemarle Street. He resumed active membership of the City Philosophical Society on Dorset Street and began to establish a reputation there as a brilliant lecturer (although its meetings were temporarily banned in 1817 under tightened sedition laws). Here he met such luminaries as Sir Richard Phillips and perhaps also John Martin, whose lifelong friend he was to become.

It is not clear how the Martins first met Faraday. At the time of his return from Europe they were still living on Marylebone High Street, a hundred yards from Dorset Street. John might well have attended the Society's meetings and heard Faraday lecturing. But it is also possible that they met through a young mutual acquaintance, Charles Wheatstone.

The Wheatstones had arrived in London from Gloucestershire in the same year as John Martin, 1806, settling on Oxford Street and then Pall Mall. They were musical instrument manufacturers and Charles' father William Wheatstone taught the flute to, among others, the wilful Princess Charlotte. Young Charles Wheatstone, born in 1802, was a gifted child with a talent for French, mathematics and physics, as well as music. He was an early devotee of bookshops, and by the age of 15 was already attempting to reproduce the experiments of famous natural philosophers, including Volta. Given his family's connection with Princess Charlotte it would not be a surprise if he was introduced to Martin, who was her drawing teacher. With Faraday (who played the flute) Wheatstone shared a love of music, of science, and of books. It is tempting to think that they met in Ribeau's bookshop over a copy of Jane Marcet's *Conversations in Chemistry*. The forming of this trio of talented mavericks was to have long-term effects on London's nineteenth-century development; in the meantime it played a minor part in events about to unfold in the northeast coalfield.

In the summer of 1815, as the industrial and social consequences of the Peace were just beginning to be felt and as the earliest locomotives puffed their way from pit-head to staithe on the banks of the River Tyne, the Society for the Prevention of Accidents in Coal Mines once more applied to Sir Humphry Davy for help. The Royal Institution forwarded the letter to Scotland, where the Davys were visiting Lord Somerville at Melrose. Davy replied on August 3rd, begging the Society to believe that he was at their disposal and promising that he would visit the coal district on his way south. At the end of the month he met the Reverend Hodgson at the Turk's Head Hotel in Newcastle where he was given all the printed material that Hodgson could obtain on ventilation and lighting in the mines – including papers relating to Dr Clanny's lamp. Hodgson then took Davy to meet John Buddle at his home in Wallsend. Here, Davy's experimental brilliance, the simple directness of his method, showed itself.

After some discussion on the nature of firedamp and its origins, about which Hodgson had his own theories, Davy took up a bucket of coal and said, 'with this I will try Mr Hodgson's experiment on coal-gas'.[8] The bucket was carried into Buddle's dining room and another bucket filled with water was sent for. Davy covered the coal with water, stirred it vigorously, then struck the coal with a heavy poker, shattering it into pieces from which bubbles could be seen emerging. Hodgson was right: firedamp was a gas inherent to coal itself. Davy then asked to see a steel mill at work in order to ascertain how much, or how little, light was needed for a miner to work by. Having satisfied himself on this matter, he left. He stayed in the coal-fields for some time and was able eventually to meet Dr Clanny and make some experiments with his lamp. There has never been any doubt that Clanny's was the first lamp.

If a large part of Davy's mind (and probably all of his wife's) wished for an end to his experimental career, it was impossible for him to turn down this challenge which would cement his reputation as the finest natural philosopher of his age. His first task was to collect enough firedamp to experiment with. Hodgson arranged for six wine bottles full of gas from a blower at Hebburn Colliery to be sent to Davy in London at the beginning of October 1815 (presumably with careful instructions as to their handling). Here, he and his former assistant Michael Faraday began to work on the gas; and here Davy made the first experimental breakthrough by determining that firedamp could not be ignited by the flame of a lamp if they were separated by an aperture of sufficiently small bore. A practicable lamp, therefore, must admit air to support the combustion of the flame through small holes at its base and emit exhaust gases through small holes above. When such a lamp was lit in the presence of firedamp, the flame enlarged and was tinged with blue but the firedamp did not explode.

October and November 1815 were months of intense activity. Davy experimented; Faraday took notes, made suggestions, drew designs and constructed lamps. Davy periodically sent word northwards, and to his scientific friends in London, that he thought he was on to something. By the end of October rash tongues, both north and south, were proclaiming the defeat of firedamp. The first public announcements came at the beginning of November at the Royal Institution and then at the Royal Society, where Davy gave an account of his and Faraday's experiments and exhibited drawings of several prototype lamps. The first public notice of Davy's discoveries in the north was in the *Newcastle Chronicle* of November 18th. At this stage Davy did not send a lamp to Hodgson to be tested in the real environment of a mine. He and Faraday took a breather, perhaps content that they had solved the basic theoretical problem. Possibly they wanted to understand the chemistry at a deeper level before producing a definitive prototype. Their most important discovery, the wire gauze, was made after the November lecture at the Royal Society, at which no mention of gauze was made.[9]

During their late autumn trials with tubes, Davy and Faraday tried a sheath made of copper gauze surrounding their naked flame and found that the heat of the flame and its exhaust gases was so dissipated by the gauze that the remaining heat was insufficient to ignite firedamp.

They then experimented to determine the optimum gauge of mesh that would give the maximum amount of light without increasing the risk of an explosion. In fact, the precise chemistry of the methane, flame and air relationship is extremely complex and has only recently been fully elucidated.[10] For Davy it did not matter. The gauze worked. Faraday or one of the other assistants at the Institution made up a new prototype lamp using a gauze sheath and Davy sent it to Hodgson in January 1816. The 'Davy' lamp was tried, with success, by a talented young engineer called Matthias Dunn at Hebburn Colliery, and pronounced a triumph.

For scientists, mining historians and engineers the precision with which these events can be pinned down was of the greatest importance for more than a hundred years. For by the time of the first trial of his lamp in a mine, Davy had a competitor. Another lamp had been tried at Killingworth Colliery on November 30th. It was designed by the colliery's resident engineer George Stephenson.

Stephenson had been at work on a lamp, entirely independent of Sir Humphry Davy, during the summer of 1815. His first two proto-types, based it seems on Clanny's original idea of isolating the flame from the surrounding atmosphere had, like Davy's, utilized small-bore tubes at the base of the lamp. The first, tested at the end of October, did not work because the exhaust gases from the flame were confined by a cap on the lamp and after a short time they extinguished the flame. A second lamp suffered the same problem. A third lamp, which incorporated apertures both above and below the flame, was ordered to be made on the 19th or 20th of November (the day after the notice of Davy's discoveries was printed in the *Newcastle Chronicle*)[11] and this was the lamp tried with success on November 30th. Whether Stephenson made use of Davy's announcement cannot be known; probably he did.

The bitter dispute which raged between Davy's and Stephenson's supporters throughout the nineteenth century now seems trite, the more so because Davy never filed a patent for his lamp and had no interest in profiting by it financially (he did not need to; his wife was wealthy). When taxed by the ever-practical John Buddle on the matter, he is said to have replied, 'No, my good friend, I never thought of such a thing; my sole object was to serve the cause of humanity; and if I have succeeded, I am amply rewarded in the gratifying recollection of having done so.'[12] This is clear evidence that Davy, sensitive to repeated attacks in the press for being a parvenu, wished to show that he was now elevated above the ranks of trade. A gentleman did not work for profit. Nevertheless, his achievement stands very high in the annals of applied science. The introduction of the Davy lamp in 1816 and its spread across the world during the next decade was by no means the end of the story, not least because it failed to prove a very good lamp in the fullness of time. But at the beginning of 1816 Davy's star, if not his flame, shone incomparably bright. He himself regarded his work on the lamp as the greatest thing he ever did.

Among Davy's fellow Prometheans the broader implications were not long in being felt. If science could be practicably harnessed to cheat death underground, what else might it not promise? Young minds of a romantic bent dreamt, as Davy himself had done, of wonderful possibilities. Could death be conquered? Could life be created?

In 1814 Percy Shelley had eloped to the Continent with the 16-year-old Mary Godwin (and her half-sister Claire Clairmont) to the fury of her father, the old revolutionary. Shelley was still married to Harriet Westbrook, later found drowned in the Serpentine River in Hyde Park. His lifestyle and libertarian philosophy were scandalous; much of his verse was unprintably radical, his ferocious energy dissipated in endless relocations, ménages and suspected murder plots. Sir Timothy Shelley having at last died and provided for Percy in his will the poet was now, in theory at least, financially independent, though he lent much too much money to Mary's father William Godwin.

In 1815 Mary gave birth to a daughter whose death two weeks later was a devastating blow. She wrote in her journal, 'Dreamt that my little baby came to life again; that it had only been cold, and that we rubbed it before the fire, and it lived. Awake and find no baby. I think about the little thing all day.'[13] In 1816, not yet married, the Shelleys were famously staying in Switzerland with Lord Byron and Claire Claremont (who was to bear Byron's child) when a competition to write a ghost story led Mary to come up with the idea of *Franken-*

stein, or The Modern Prometheus. In her 1831 introduction to a new edition of *Frankenstein* Mary Shelley cited a number of late-night conversations between Shelley and Byron during which they discussed the philosophical and poetic consequences of science. Erasmus Darwin, grandfather of Charles and leading light of the Lunar Society of the late eighteenth century had, they recalled, preserved vermicelli in a glass until it spontaneously generated new life. Shelley, as a keen amateur scientist, had attended lectures at the Royal Institution by Davy and others. Mary had met Davy socially because he was a friend of her father's. News of the invention of the miners' lamp had probably reached them. It is not difficult to see something of Davy, the charismatic magus of chemistry and electricity, in the tragic figure of Victor Frankenstein. But Frankenstein is not Davy. Davy had himself warned of the dangers of mere 'speculative' philosophy, and more likely appears as Frankenstein's unsuspecting mentor Professor Waldman.

Victor Frankenstein is at least as much an embodiment of the evils of 'Pandora's pent-up mischief' of which Mary's own mother had warned in 1792.[14] He is, as Shelley herself recognized in her title, the dark side of Prometheus, so evocative a figure that he can instantly be recognized in such modern iconic figures as the Darth Vader of *Star Wars*, the robotic parents of the *Terminator* and the replicant-builders of *Blade Runner*. His rejection of his own 'child' offers echoes of Mary's relationship with her father, and has been recognized by many critics as a reflection of the God–Adam relationship in *Paradise Lost*. He could equally be a personification of her husband Percy, soon to establish himself as the chief prophet of the cult of Prometheanism. He, perhaps more than any of his contemporaries, embodied the talent for self-destructive brilliance that defines his age.

On their return to London at the end of 1816 Shelley, honoured by Leigh Hunt in his *Young Poets*, met John Keats for the first time and became part of north London's radical set, while Mary worked on her novel. Shortly thereafter Shelley may have met John Martin. He paid his fellow Promethean a poetic compliment in *The Revolt of Islam* (1817) with the lines:

the King, with gathered brow, and lips
Wreathed by long scorn, did idly sneer
With hue like that when some great painter dips
His pencil in the gloom of earthquake and eclipse.[15]

Frankenstein was published anonymously in 1818 and greeted with shock, cynicism and admiration in almost equal measure. At least one critic correctly identified the authoress, admitting that her writing bore 'marks of considerable power … but this power is so abused and perverted, that we should almost prefer imbecility'.[16] Disgust was a common reaction among reviewers. One called it 'a tissue of horrible and disgusting absurdity'.[17] The *Belle Assemblée* was more sympathetic, if cautious:

> This is a very *bold* fiction; and, did not the author, in a short Preface, make a kind of apology, we should almost pronounce it to be *impious*. We hope, however, the writer had the moral in view which we are desirous of drawing from it, that the *presumptive* works of man must be frightful, vile, and horrible; ending only in discomfort and misery to himself. [18]

The most intelligent and perceptive notice was Walter Scott's in *Blackwood's Edinburgh Magazine*. He mistakenly believed *Frankenstein* to have been written by Percy Shelley. In words which might equally have applied to John Martin's pictures, he recognized that despite the fantastical caprice of the novel's premise, it belonged legitimately to a class of fiction in which: 'the laws of nature are represented as altered, not for the purpose of pampering the imagination with wonders, but in order to shew the probable effect which the supposed miracles would produce on those who witnessed them'.[19] Scott was careful to analyse the story's literary merit, addressing its evident flaws. But he could not help admiring the author's Promethean spirit: 'The author seems to us to disclose uncommon powers of poetic imagination. The feeling with which we perused the unexpected and fearful, yet, allowing the possibility of the event, very natural conclusion of Frankenstein's experiment, shook a little even our firm nerves …'[20]

Within a year of its publication the horror of events in Manchester lent to the moral imperative of *Frankenstein* an almost prophetic sense of doom. But the year 1818 was itself not without events of Promethean irony. It was the year in which Karl Marx was born and Jonathan Martin was incarcerated for threatening to shoot Dr Legge, the Bishop of Oxford; in which an explosion at Wallsend Colliery killed four men; and in which William Martin constructed his miners' safety lamp.

Since his retirement from the sea Jonathan Martin had been living quietly, so far as one can tell, in Norton, near Stockton-on-Tees. He found work as a tanner and married. In 1814 his wife Martha bore them a son whom he named Richard after his favourite brother. After the death of Fenwick and Susan Martin the previous year Jonathan had a dream in which his mother and sister appeared to him and foretold that he would be hanged. He felt this was a sign that his youthful misdemeanours had not been forgiven by God. In reparation he became a frequent attendant at both church and chapel and eventually joined a Methodist congregation to the apparent disgust of his wife. In an increasingly desperate state of religious confusion he attempted to preach a sermon in the ranting manner of the Methodists, to an Orthodox church congregation from which he was dragged by the church clerk. He managed to alienate friends, colleagues and fellow Methodists alike with his erratic behaviour, and was eventually sacked by his employer.

After a short time being supported by the poor rate, Jonathan and his wife and child sought work further afield, first in Whitby and then at Bishop Auckland where they stayed for several months in 1817 or early 1818. Here he began to acquire a reputation for disrupting church services and was taken before the local magistrate who, having read a supporting letter from his new employer, acquitted him. It was now that he saw an announcement that Dr Legge, the Bishop of Oxford, was to hold a confirmation at Stockton. Having heard that the bishop was a good man Jonathan thought he would like to confirm it for himself. On a recent visit to William in Newcastle his brother had given Jonathan an old pistol from his militia days. Its barrel was broken and it would have blown his own hand off had he tried to fire it, but he told his wife he thought he might shoot the bishop as a test of the clergyman's faith. His wife probably then hid the gun, for it disappeared.

At the service the wonderful size of the bishop's stomach con-firmed Jonathan in his belief that the gentleman had drunk a great deal of wine. Jonathan did nothing at all to threaten the good cleric, but his wife must have warned the vicar, who questioned him. Jonathan admitted that he thought all clergymen ought to be shot. This time, brought before the magistrates, he was asked if he really would have shot the bishop and he said he might have done if the bishop had not answered his questions in a satisfactory manner. That was enough for the magistrates. They ordered him to be confined in a madhouse.[21]

If William Martin regretted lending his younger brother a pistol he did not record the fact in any of his memoirs; probably he was wrapped up in his own obsessions. He had returned to Tyneside in 1814 after the disbanding of the Northumberland Militia. Shortly afterwards he married a woman whose name he fails to record in his autobiographical tracts. She was, he wrote, an 'inoffensive' woman – a grudging compliment to a dressmaker of local renown who employed no less than 60 apprentices in her years of business and kept William for the rest of her life.

By the time of Jonathan's incarceration William and his wife had probably moved to Church Bank, Wallsend. With his lifelong interest in the mines William could not fail to have followed the fortunes of John Buddle, George Stephenson and the lamp saga. The innovations being introduced in the Tyne Basin coalfield were already attracting interest from much further afield. In 1816, the year in which Davy's and Stephenson's lamps were introduced, Grand Duke Nicholas of Russia, the future Tsar, visited Newcastle. Here, as well as being shown Davy's new lamp, he saw George Stephenson's engines at Killingworth Colliery and a locomotive called the *Steam Elephant* in use at Wallsend designed by John Buddle and William Chapman.[22] The significance of this engine was that it introduced Chapman's idea of distributing an engine's weight across two 'bogies', the four-wheel suspension units which all modern trains employ to enable them to go round bends without being derailed. At Wallsend, Buddle showed the Grand Duke around his operations but the duke declined to descend the pit itself, apparently shrinking back from the shaft crying (in French), 'Ah my God it is the mouth of Hell, none but a madmen would venture in it.'[23]

With the invention of the lamp and the locomotive, Tyneside was now at the forefront of industrial progress. Within a few years America and Russia were taking more than a passing interest in the technical revolution taking place here. Both countries later sent confidential agents to learn its secrets. All was not entirely well, though. Thousands of seamen discharged from the navy had returned and were trying to force local ship owners to take them on. Other innovations threatened existing jobs: wooden chutes mounted on staithes[24] were being built along the river so that coal wagons could dump their loads directly into colliers; as a result the number of keelmen employed on the River Tyne was falling dramatically. With the militias now largely disbanded, militant workers once again felt free to flex their industrial muscle.

William Martin saw all this. He also saw the new safety lamps and straightaway envisioned improvements to them. He told George Stephenson that his lamp was worse than a candle and would easily break if dropped. Davy's he saw would fire if the gauze got too hot. So he set about constructing his own lamp and had it taken to Wallsend Colliery. John Buddle admitted to him that his lamp was a good one, but that it would be better with Davy's gauze inside the glass. William made a second lamp which incorporated a gauze sheath and, according to his version of the story, took it to Alnwick Castle to show the Duke of Northumberland. It was, he remembered, on his way back that he heard of an explosion at the Church Pit in Wallsend.

This explosion happened in August 1818. According to William, he met one of the miners, William Thompson, who had witnessed the blast that killed four men. 'The lamps!' he shouted to William. 'What do they mean by calling them safe when they aren't? I saw the lamp fire in the lad's hand.'[25] It was true: Davy's lamp was not safe. It was, to begin with, open to abuse. Miners would lift the gauze off for a better flame or to light their tobacco pipes with. The gauze was easily damaged, either by accident or through long use when repeated heating and cooling weakened it.

It is a salutary truth that deaths in mine explosions increased after

NEW PRINCIPLES, or the MARCH of INVENTION.

The march of the Industrial Revolution caricatured (above) and a dire warning
(below) of the evils unleashed from Pandora's jar: the execution of Louis XVI of France.

FOUR PROMETHEANS

TOP LEFT Percy Bysshe Shelley: poet, revolutionary and atheist.

TOP RIGHT Caroline Norton: poetess, wronged wife and mother and feminist pioneer.

ABOVE RIGHT William Godwin: the old anarchist; Percy Shelley's father-in-law.

LEFT William Martin: eldest of the Martin brothers. Eccentric inventor of a miners' safety lamp, cartoonist, doggerel poet, athlete, soldier and self-styled Philosophical Conqueror of all Nations.

BELOW Jonathan Martin: after a traumatic childhood filled with visions of fire and brimstone, his inadvertent career in the navy was followed by a lapse into religious mania and a tragic attempt to destroy York Minster.

LEFT John Martin: small, neat, passionate and loyal. Artist and amateur engineer, he single-handedly invented, mastered and exhausted an entire genre of painting – the apocalyptic sublime.

ABOVE *Belshazzar's Feast* (1821):
John Martin's defining work, a
cinemascope epic on a vast scale
and a seditious parody of
George IV's extravagance.

RIGHT *The Bard* (1817): with
echoes of Bruegel and inspired
by a Thomas Gray ode, John
Martin's immense canvas was
nevertheless a public statement
in praise of free speech.

ABOVE *London's Overthrow* (1830): Jonathan Martin's manic drawing from his early days in Bedlam; a dire warning to the inhabitants of 'Babylon-on-Thames'.

LEFT *Sadak in Search of the Waters of Oblivion* (1812) was John Martin's first commercial success, laying a marker for his iconoclastic career. Sadak, like Martin, is the outcast fighting overwhelming odds.

OPPOSITE PAGE RIGHT
Wheatstone's and
Cooke's five-needle
telegraph (1837):
a Promethean
triumph of science,
engineering and
entrepreneurial spirit.

RIGHT Marc Brunel:
French émigré
visionary and
engineer, whose
impressive legacy to
Britain included his
son, Isambard
Kingdom Brunel.

FAR RIGHT Joseph
Bramah's 'unpickable
lock' of 1784: his
problem was to find
someone who could
replicate its miniature
perfection on a
commercial scale.

ABOVE Humphry Davy:
dazzling superstar of the
Royal Institution, the man
who made science compatible
with gentility. His safety
lamp (LEFT) is a supreme
icon of the Promethean spirit.

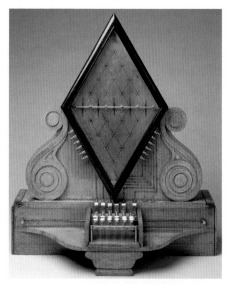

The Artist
who can make an
Instrument that will pick
or Open this Lock, shall
Receive 200 Guineas
The Moment it is
produced.
Applications in Writing only.
Bramah's PATENT Locks.
CAUTION.
The Public is Respectfully Informed
that every LOCK, made by
BRAMAH & Co is Stamped
with their Address
124 Piccadilly

ABOVE Charles Babbage: the
man responsible for conceiving
the digital computer a hundred
years and more before its time;
and (LEFT) his Difference
Engine Number One.

LEFT J.M.W. Turner's self-portrait, completed in 1798.

BELOW *Rain, Steam and Speed – The Great Western Railway* (before 1844): Turner's unparalleled achievement was to chronicle with sensitivity, but without judgement, the passing of the old age and the blurring onward rush of modernity. By the 1840s mere objects had ceased to interest him; many thought he had gone mad. He had begun to paint with pure light.

the introduction of the Davy lamp. In 1819, 2 men and 35 boys died at Sheriff Hill Colliery and 13 died at George Pit, Lumley. In 1820, 2 died at Jarrow Colliery and 52 more were killed by a blast at Buddle's own colliery at Wallsend; 6 more died at Felling ... and so on.

According to William Martin's account (which must be treated with caution because his bitterness over the lamp was a cause célèbre for the rest of his life) when he interviewed William Thompson later the miner said he must not speak about it or he would lose his job. This is plausible. Sir Humphry Davy had selflessly invented a lamp which had been lauded by all the engineers in the coalfield; men were now making money manufacturing it and miners had to trust it.

Precisely what role John Martin played in the lamp saga is not clear. The Sezincote sketchbook of 1818 contains two sketches of a miners' safety lamp. One of them is a rough trial, the other a detailed drawing. It has been supposed by John's biographers that these are drawings of William's lamp. But if that was the case, William would surely have asked John to make engineering plans for a tinsmith to work up from his own sketches. Perhaps such drawings existed. But the Sezincote sketches look as if they are John's own thoughts roughed out. It is certainly an interesting design. There is a solid round base with feet so that it may stand up alone, as well as a ring at the top for carrying. Above the oil reservoir is a necklace of tiny apertures for admitting air, and a filler cap. The body of the lamp containing the flame looks as though it might be gauze, though this is not absolutely clear. It is protected by struts which lead from the oil reservoir to a cap which looks as if it must be of glass. In the light of the subsequent development of the lamp, notably by the neglected William Clanny, this arrangement is upside down: the glass should surround the flame with the gauze sheathing the cap. Its limitations would have been evident in a worked-up model.

William's original lamp does not survive so one cannot be sure whether or not John's drawing was done for and with William. But it is possible that it was made in response to conversations with John's friends Faraday and Wheatstone when William's involvement became known to him. It may be, in other words, an entirely independent design. If so, it was never constructed. The possibility exists, though, that Michael Faraday, whose knowledge of Davy's thinking on the lamp was intimate and who may himself have constructed the prototype Davy lamp, influenced William's design directly or indirectly via John.

William's second lamp had a short but significant history. He succeeded in persuading the Duke of Northumberland to intervene on his behalf and in 1819 his lamp was put through its paces at Willington Colliery. Understandably, William kept the affidavit drawn up for him there. Written on 19 April 1819, it is quoted here in full:

Gentlemen, – These are to certify that Mr Martin's lamp has been tried with Sir Humphry Davy's, by order of John Watson, Esq. and Mr Johnson, viewers of Willington colliery; we, the undersigned, were strictly charged by those gentlemen to act impartially, and to give no reports on their comparative merits, but such a one as we could substantiate on oath. We took the lamps down the pit, and giving each a good wick, we lighted them. And noted the time of lighting them by a watch. In one half-hour Sir Humphry Davy's lamp was entirely filled up with smoke and soot. Mr Martin's lamp burned five-and-a-half hours, and did not appear dull, for by blowing at the top of the lamp, it became as brilliant as when newly lighted. Sir Humphry's lamp is much superior to Mr Stephenson's, which cannot conscientiously be called a safety lamp; but Mr Martin's surpasses Sir Humphry's by the duration of its burning and the brightness of its light, and cannot fail of giving satisfaction to both masters and workmen. The merit of Mr Martin's lies in the chimney and small ventilators; it stands proof against fire in every dangerous part of the colliery; and is looked upon to be much cheaper, on account of its durability, and much safer on account of its top where the chimney is. Where Sir Humphry's is the weakest, Mr Martin's is the strongest; in short, Mr Martin's lamp is not exceeded either in light or safety, and is more beneficial, both to masters and men, than any lamp that has been introduced to the coal-mines. We do hereby certify, that we have justly tried Mr Martin's well-designed lamp, and that we have not enlarged on its merits more than truth demands of us; indeed we cannot find words to express its superiority. We calculate that one of Mr Martin's gauze-wires will outlast ten of Sir Humphry Davy's.

    Clem. Simson

    Joh. Gaskin

    Andrew Bell

    Demr. Hornsby [26]

William's sense of outraged pride that his lamp was not adopted after this endorsement is understandable. It rose to a fury a few months later when he visited the manufactory of John Watson on High Bridge in Newcastle and found George Stephenson there, having a copy of his (William's) lamp made. According to his own recollections, William rounded on Stephenson, rapped his cane on the table and cried, 'So you've made a botch and a clumsy exhibition of your own lamp, and now you'll make a seizure of mine!'

Stephenson's 'Geordie' lamp was widely adopted in the coal mines of Newcastle as word spread of the deficiencies of the Davy lamp. It is not beyond understanding, given what is known of George Stephenson's career, that his biographers make no mention of William Martin's lamp. It is possible that William's design in fact survives in the 'Geordie' lamp which was adopted in so many pits across the region. Stephenson was fêted as Davy had been. William never received a penny royalty. For Dr Clanny, who made numerous improvements to his safety lamp over the years, recognition did not come until 1848 when he was presented with a silver salver at the Athenaeum in Sunderland.[27]

The saga of the miners' safety lamp is not quite moribund. Even today, a safety lamp must be carried into a mine every day, for its flame acts as a sensitive indicator of the presence of methane and carbon dioxide. And there is a certain irony in the fact that the Olympic torch, commemorating Prometheus's original theft of fire, is still carried between Olympia and its quadrennial destination by means of a miners' safety lamp.

CHAPTER NINE

Belshazzar's Feast

'I'VE BEEN THREE TIMES RUNNING to see the exhibition in Pall Mall to admire The Capture of Babylon by Martin. He adds the greatest distinction to contemporary art. Oh, what a sublime thing!'[1] William Beckford, owner of the Gothic Revival fantasy Fonthill Abbey and patron of J. M. W. Turner among others, was writing at the beginning of February 1819. Once again John Martin had risked reputation and finances alike in executing a subject he thought the public would respond to. Once again he was proved right. *Babylon* was conceived on an even grander scale than *Joshua*, measuring 61 by 96 inches. It was a dazzling display of architectural perspective and atmospheric lighting which showed Martin's increasing confidence in mastering a genre that he himself was in the process of creating. There are the trademark huddled, diminished figures; the bolt of lightning from a furious sky; the fantastic oversized buildings of an ancient world (Tower of Babel and all) which existed only in the minds of Bible-readers and their Geordie high-priest. Escaping from the dank chill of a miserable London February, crowds came in large numbers to see this new wonder (its present whereabouts are unknown). Among other gratifying reviews *The Examiner*, whose art critic was another Hunt brother, Robert, was almost reckless in its praise: 'The spectators crowd around it, some with silence, some with exclamatory admiration; sometimes very near to look at the numerous small objects that cannot be distinguished at a distance, sometimes farther off to feast upon the grandeur of the whole; leaving it, but coming back to look at it again ...'[2]

Aside from its vivid recreation of an Old Testament narrative that would have been perfectly familiar to the churchgoing public, *Babylon* had a sensational three-dimensional effect on its audience. Its closest modern equivalents might be surround-sound and the giant IMAX screen; its impact on a public who had never seen a photograph or a moving image must have been greater still. Unlike the static posed serenity of the historical paintings approved by the Royal Academy, it came almost alive with its sense of scale and perspective, full of visual trickery and movement. When it was later shown at Martin's Egyptian Hall exhibition in 1822 it was accompanied by a catalogue description of all the main architectural and narrative features, fully referenced by the historical sources Martin had consulted. Such catalogues became a standard feature of his works from then on, giving the impression that they were not just art, but historical exhibits that required explanation.

For the first time Martin had chosen the British Institution rather than the Academy for a painting's first major showing. The Institution was suitably repaid by the gratifying numbers who flooded into that year's exhibition and bought more than a thousand pounds worth of paintings. *Babylon* was snapped up by a well-known collector, Henry Hope, at the full asking price of 400 guineas. For the first time in his life John Martin found himself unencumbered by debt, having borrowed the money for the house in Allsop Terrace from William Manning, the MP who had bought *Sadak* and whose offer of financial help John had reluctantly taken up. Manning was repaid immediately. John was overwhelmed by ...

> ... a sum which was enough to set me free, to unmanacle me
> from the chains of debt, to place me above want ... If I had not
> succeeded now, I must have sunk. Increased rent, increased
> expenses, every way. I felt as much joy and glut of delight, at
> painting a picture on which I put four hundred guineas and getting
> it, as Wellington must have felt in conquering Bonaparte. I had
> thirsted to succeed as a painter. Was this not success?[3]

John Martin might also have indulged in a wry smile when he thought of his former colleagues at William Collins's, languishing on two pounds per week. It would be easy to view *The Fall of Babylon*, its correct title, as the proselytizing raving of a fundamentalist (in later years John was very often, and understandably, confused with his brother Jonathan and known as 'Mad' John Martin) if it were not for its very obvious contemporary secular message. For Babylon, read London.[4] John Martin may have been cashing in on a vogue for shocking images and for historical reconstructions of mysterious ruins from contemporary travellers' tales, but he was also tapping into the highly charged prevailing political atmosphere. Unemployment, high prices, rioting and the Gag Acts of 1817, the death of Princess Charlotte and the extreme unpopularity of Lord Liverpool's Government (let alone that of the Prince Regent): all seemed to presage revolution. In portraying the glory of civilizations now reduced to dust Martin was not just satirizing the establishment: he was warning them.

There is much more than a passing resemblance between Martin's *Fall of Babylon* and Percy Shelley's 'Ozymandias' (1817):

> I MET a traveller from an antique land,
> Who said: Two vast and trunkless legs of stone
> Stand in the desert … Near them on the sand,
> Half sunk, a shattered visage lies, whose frown,
> And wrinkled lip and sneer of cold command,
> Tell that its sculptor well those passions read
> Which yet survive, stamped on these lifeless things,
> The hand that mock'd them and the heart that fed:
> And on the pedestal these words appear:
> 'My name is Ozymandias, king of kings:
> Look on my works, ye Mighty, and despair!'
> Nothing beside remains. Round the decay
> Of that colossal wreck, boundless and bare,
> The lone and level sands stretch far away.[5]

That Martin and Shelley were perfectly in tune in drawing on grandiose ancient ruins to score a contemporary political point can perhaps be attributed to their mutual friendship with the Hunts and their admiration for William Godwin, as much as to the vogue for all things Middle Eastern. It may also owe much to the prevailing Promethean spirit among the reformist, self-made talents of north London who increasingly felt it their duty to challenge a defensive, paranoid elite that still ruled in almost complete denial of the tide lapping at their feet. The Prometheans did not, in early 1819, feel that they had to incite revolution. They had merely to portray what they believed to be London's fate to make their point. But their mood was about to change.

Far away from Babylon-on-Thames, on the moors above the spinning and weaving towns of northwest England, militant workers, half-starved many of them and dressed in rags, drilled unarmed in orderly fashion under the eyes of magistrates and government agents-provocateurs. Their purpose was to argue for change, to petition the government, and by sheer force of numbers to gain for themselves the representation they were denied by an iniquitous electoral system. They were campaigners, not revolutionaries. Their message may have lacked the artistic irony of Shelley and Martin but it was nevertheless unmistakeable.

One weaver, Samuel Bamford of Middleton in Lancashire, found himself and several others arrested and accused of high treason for having held a private meeting to discuss parliamentary reform. Brought to London, he spent a nervous, though surprisingly comfortable, spell in Cold Bath Fields prison. From here he was periodically taken before the Home Secretary, Lord Sidmouth, to answer for himself.

It says much about the times that after four interviews Bamford's protestations of innocence were accepted and he was released in time to witness events in Manchester later in the year (although several of his friends remained in prison). It says more, perhaps, about the lot of Britain's postwar labourers that the regime at Cold Bath Fields, harsh as it was, seemed quite comfortable to Lancashire weavers who, while they were in prison, at least ate regularly and were given fresh linen to wear.

Reformers in London were not unaware of events going on in the industrial towns of the north. Increasingly, the Royal Mail was being used to deliver regional and national newspapers and pamphlets around the country. Despite increases in stamp duty these were read in great numbers at clubs and subscription libraries. At the beginning of 1819 there was an almost tangible anticipation of imminent change, for good or ill.

The apocalyptic warning to the capital issued by John Martin in February chimed with another, more direct message which appeared in *The Examiner* in January from the pen of fellow Promethean Leigh Hunt: 'This is the commencement, if we are not much mistaken, of one of the most important years that have been seen for a long while. It is quiet, it seems peaceable … but a spirit is abroad, stronger than kings, or armies, or all the most predominant shapes of prejudice or force …'[6] Popular discontent with the postwar economic crisis and with a lack of parliamentary reform was made worse by an unsatisfactory result in the 1818 general election. Not only did the Whig opposition not gain materially in members, but its leadership was in disarray. Its most promising leaders in the Commons, Sam Whitbread and Samuel Romilly had both committed suicide and its shining light, Henry Brougham, gave the impression of being an unprincipled chancer.

The landowning Whig nobility, at least until late in 1819, had managed to convince themselves (they must have been deaf and blind) that there was no real appetite for reform in the country. Strategically, their backing for the campaign to emancipate Catholics was a mistake: just and proper as it was, the campaign had little support in the country as a whole. They must also still have been smarting from disappointment with their erstwhile 'friend' the Prince Regent, although if they were it was their own fault, as William Makepeace Thackeray later pointed out:

> The friendship between the Prince and the Whig chiefs was impossible. They were hypocrites in pretending to respect him, and if he broke the hollow compact between them, who shall blame him? His natural companions were dandies and parasites. He could

> talk to a tailor or a cook; but, as the equal of great statesmen, to set
> up a creature, lazy, weak, indolent, besotted, of monstrous vanity,
> and levity incurable – it is absurd.[7]

Apart from his reform promises, the Prince Regent's greater political sin was to have failed in his duty to the Hanoverian succession. Having lost his daughter Charlotte, and having been estranged from his wife Caroline since a fortnight after their marriage in 1795, he was most unlikely to produce an heir. The divorce question had been raised many times but the prospect of this couple's very dirty linen being aired in public, while it was sure to enrapture the populace, was politically out of the question, at least while the old king still clung to life. George III's second son Frederick, Duke of York, would die childless. William, Duke of Clarence, was to have two children but both of them died in infancy. The hope, the very faint hope, was that Edward, Duke of Kent (George III's fourth son) would provide an heir. He had married Mary of Saxe-Coburg Saalfeld (Prince Leopold's sister) in 1818 as an act of duty, having lived in happy sin with Mme Laurent for more than twenty years.

In the spring of 1819 Mary was expecting her first, and as it turned out only, child. On May 24th Princess Victoria was born. It was good news for the government and the Hanoverian dynasty; even so, the likelihood that she would come of age before her father (dead in 1820) or her uncles Frederick (1827), George (1830) and William (1837) passed on seemed most unlikely at the time. In the meantime the majority stakeholder in this young princess was her 'Dearest Uncle' Leopold who, having been robbed of the chance to sit in the consort's throne himself, now saw his stock rising as a potential regent to his niece.

Spring's bloom was short-lived. It turned into a hot summer of unrest. A Promethean comet appeared in the sky. In July 1819 seventy thousand people assembled at a meeting in Smithfield calling for parliamentary reform. That same month the magistrates of the manufacturing town of Manchester which, with a population of more than a hundred thousand, still returned no member to Parliament, wrote to Lord Sidmouth warning that the deep distresses of the workers, oppressed with hunger, led the authorities to believe that a general rising was being planned. Later that month a committee 'To strengthen the Civil Power' met at the Manchester Police Office and resolved to request government to furnish arms and accoutrements for 1,000 men.[8]

A public meeting planned by the recently formed Patriotic Union Society for 9 August became the focus of the magistrates' fears. The famous Radical Henry 'Orator' Hunt had been invited to address the meeting at St Peter's Field to call for annual parliaments and 'universal' male suffrage. It was not just the magistrates who were concerned for public order. The Radicals, and Hunt in particular, were adamant that no provocation for violence was to be given to the authorities. The drilling of companies on the moors above the towns of Lancashire was in truth designed to insure against violence. Arms, weapons of any type, were forbidden. Men and women would wear sprigs of laurel in their hats, their best Sunday shirts and clean neckerchiefs. There were to be bands, dancing. Each town was to carry its own colours, some of blue silk with inscriptions in gold letters: 'UNITY AND STRENGTH', 'LIBERTY AND FRATERNITY', and so on. Hunt, Bamford and others stressed that if the authorities came to arrest any member of the meeting they should be suffered to conduct their lawful business. There was to be no retaliation.

The meeting arranged for the 9th of August was banned by the authorities, so the Radicals rearranged it for a week later. The nervousness of the magistrates had by now been heightened to a pitch by reports from infiltrators that some of the Radicals intended to use the meeting to ignite a country-wide Jacobin revolution. Some of them no doubt wished for just such an outcome but the organizers believed that their careful planning would prevent trouble from flaring. Radicals were by now well used to spotting government and police agents, and many of these were dealt rough justice with clubs and clogs in the back streets of small towns.

By the time of Henry Hunt's arrival at the platform in St Peter's Field at least sixty thousand men and women (it might have been many more) had assembled in a vast throng. The various town bands were playing, flags were flying; the Radicals' most ambitious hopes for the size of the crowd had been justified. Here was an open, honest plea for change from the people. Who now could claim that there was no appetite for reform? Hunt made his way to the front of the platform, took off his white hat (the symbol of radicalism), and addressed the people. Samuel Bamford, having heard Hunt many times before and being satisfied with the peaceable nature of the meeting, had determined to withdraw for refreshment when he heard a disturbance in the crowd and made his way to where a line of cavalry could be seen, harnesses clanking, horses champing, sabres drawn. The crowd cheered them. The cavalry raised their sabres high, as if in salute. Then, to Bamford's horror,

> striking spur into their steeds, they dashed forward, and began cutting the people.
>
> 'Stand fast,' I said, 'they are riding upon us, stand fast.' The cavalry were in confusion: they evidently could not, with all the weight of man and horse, penetrate that compact mass of human beings; and their sabres were plied to hew a way through naked held-up hands, and defenceless heads; and then chopped limbs, and wound-gaping skulls were seen; and groans and cries were mingled with that horrid confusion.[9]

The butcher's bill for that day was eleven killed, a hundred injured by sword-cuts and several hundred more crushed in the panic which followed the initial charge. It was England's Tiananmen Square. Hunt, Bamford and others were arrested, tried by Special Jury and imprisoned on evidence which, even at the time, reeked of perjury. Neither Manchester's Deputy Constable, Joseph Nadin, nor any member of the yeomanry who had wielded swords, bayonets and muskets, were charged with any offence. Lord Sidmouth wrote to congratulate the magistrates on their handling of this dangerous insurrection.

News of what the *Manchester Observer* christened the Peterloo Massacre scandalized the literate liberal classes and radicalized workers across the country. The *Manchester Guardian* was born from their indignation. Henry Brougham wrote angry letters to his political friends about the state of the judiciary, but the parliamentary recess of that summer saved the government from an immediate political inquest. In November 1819 Lord Sidmouth drew up plans for emergency legislation in the face of what looked like perfect conditions for revolution. By the end of December the so-called Six Acts had been passed. The drilling of bodies of men, armed or unarmed, was forbidden. New powers were given to magistrates to seize arms (many men had kept their service muskets from the war). Stricter laws on blasphemy and sedition and on private gatherings, new powers of search and rules for juryless trials were compounded by a punitive increase in stamp duty which was intended to prevent the poorer classes from reading newspapers. The Royal Mail no longer distributed newspapers across the country.

News of the massacre spread across Europe with great speed. J. M. W. Turner and Humphry Davy (sharing a love of Italian art and fishing), the poet Tom Moore (a debt exile), Sir Thomas Lawrence (artist and former lover of Queen Caroline) and the sculptor Sir Francis Chantrey, all of whom met in Rome that autumn, had heard by the beginning of September. The same month Percy Shelley, the revolutionary in exile, wrote to his London publisher Charles Ollier from Leghorn (Livorno) that a torrent of indignation was boiling in his veins.[10]

If he had been in England, if he had been a more practical than theoretical visionary, if anyone had been prepared to publish his response, Shelley might have made more of an impact, might have acted as a touchstone for the Radicals. As it was he vented his spleen in 91 verses in *The Masque of Anarchy*, a burst of poetic vitriol so obviously seditious that no one, not even Leigh Hunt, would publish it. It was self-indulgence from a prophet too far away, too abstracted, to be heard by the people for whose cause he claimed to be fighting. Infinitely more effective, had it been published, was his controlled and deadly indictment of a rotting society in 'Sonnet: England in 1819':

> An old, mad, blind, despised, and dying king,
> Princes, the dregs of their dull race, who flow

> Through public scorn, – mud from a muddy spring,
> Rulers who neither see, nor feel, nor know,
> But leech-like to their fainting country cling,
> Till they drop, blind in blood, without a blow,
> A people starved and stabbed in the untilled field,
> An army, which liberticide and prey
> Makes as a two-edged sword to all who wield,
> Golden and sanguine laws which tempt and slay;
> Religion Christless, Godless – a book sealed;
> A Senate, – Time's worst statute unrepealed,
> Are graves, from which a glorious Phantom may
> Burst, to illumine our tempestuous day.[11]

If Shelley wanted to be heard back home by more than his close friends he had to play a subtler hand and issue a call to arms to his fellow Protheans. Now he determined to construct a thesis in verse. Between his arrival in Italy, torpid and consumptive in 1818, and 1820 when it was published, Shelley created the lyrical drama *Prometheus Unbound*. The play narrates Prometheus' awakening to his terrible isolation and torture, his release by Heracles and his promise to fulfil his commitment to man in the fight against tyranny. At the Baths of Caracalla in Rome in the spring of 1819 Shelley wrote the first three acts; the fourth was finished in Florence at the end of the year. His aim was not to complete Aeschylus' lost work[12] but to rework the myth as a poetic statement of his own political philosophy. According to Mary's later notes[13] Shelley's theory of the destiny of humanity was that it was not inherently evil, that it was perfectible. This echoes the views of his father-in-law William Godwin in *An Enquiry*, which had convinced Shelley of a more serious purpose in life than mere liberty.

Shelley's Prometheus, unlike his wife's portrayal of the Titan in *Frankenstein*, is virtuous but wronged, his theft of fire an unavoidable act of moral imperative. He gives man speech and speech creates thought, 'which is the measure of the universe'.[14] Prometheus, like Christ, suffers the ultimate penalties of torture and humiliation for these gifts and 'hangs withering in pain'.[15]

The setting may be mythical but the politics are entirely contemporary. The chorus encourages its audience to look 'where round the wide horizon many a million-peopled city vomits smoke in the bright air' and 'mark that outcry of despair'.[16] Prometheus' own blood, his 'drops of agony' hold the promise for mankind that a disenchanted nation shall 'spring like day from desolation'.[17] Shelley's hope, voiced by his Chorus of Spirits, is that they will build …

> In the void's loose field
> A world for the Spirit of Wisdom to wield;
> We will take our plan
> From the new world of man,
> And our work shall be called the Promethean.[18]

If there was any doubt that this work represented a manifesto for the Prometheans, Shelley removes it in his preface. Comparing Prometheus with the anti-hero of *Paradise Lost*, he argues that Milton's Satan, for all his courage, majesty and opposition to omnipotence, was tainted with 'ambition, envy, revenge and a desire for personal aggrandizement'. Prometheus, in contrast, 'is, as it were, the type of the highest perfection of moral and intellectual nature impelled by the purest and truest motives to the best and noblest ends'.

No clearer statement could have been made by any contemporary of the ideology shared by the acolytes of the cult of Prometheanism, with Shelley as its prophet, John Martin its high priest and their friends and colleagues its proselytes. Shelley, like both Martin and Hunt instinctively recognizing the year 1819 as a low point in the fortunes of the English people (and perhaps with an eye on his own fragile destiny) issued in his preface a rallying cry to his fellow Prometheans: 'The great writers of our own age are, we have reason to believe, the companions and forerunners of some imagined change in our social condition or the opinions which cement it. The cloud of mind is discharging its collected lightning, and the equilibrium between institutions and opinions is now restoring or is about to be restored.' For writers read also painters, scientists, engineers and educationalists. The optimism of Promethean perfectibility struck chords with poets envisioning heaven on earth; with scientists exploring the potential benefits of chemistry and physics; with engineers experimenting on new materials and machines and with artists exploring new media. To what extent they were aware of Shelley's call to arms is difficult to say, but if he was the Prometheans' most articulate voice, he was not the only one, as the following years showed. Meanwhile, the Olympian British establishment, who had most to lose, remained in a state of nervous denial, gagging public dissent but impotent to stem the reformist tide and with no clear idea how to reconcile the opposed interests of landowners and labourers.

Now that John Martin was a man of means, or at least of the two hundred pounds left to him after repaying his debt to Manning, the household on Allsop Terrace became a centre for social gatherings. At first, people had come to watch the Martins and the Hunts at chess. Others came to be seen and to chatter. Gradually chess became a sideline to singing, laughing and the passionate discussion of politics, arts and science. The house, already filled with children, became a lively stage playing host to intellectuals, artists, scientists and engineers.

The guests were generally but by no means always of a sympathetic political persuasion. John's son Leopold recalled many years later a morning's stroll with his father during which they called on his friend Edwin Landseer, the painter of magnificent animals, who lived not far away at Number 1, St John's Wood Road. They found the painter 'much dejected, quite out of spirits and sorts'. Their strong suspicion was that he had stayed late at the gaming tables and lost a great deal of money, as was his wont. They managed at length to induce him to go for a walk with them. In the park they passed a lady to whom John Martin raised his hat in greeting. This was Mrs Fonblanque, wife of John's friend Albany Fonblanque who was a radical journalist and later Leigh Hunt's successor as editor of *The Examiner*. Landseer was shocked. 'Why,' he exclaimed, 'her husband is a radical! How fortunate that you were not seen by more than a true friend, one not likely to expose you.'[19] Returning to Landseer's house they found their friend the Reverend Sydney Smith, one of England's great wits.

Landseer asked Smith if he would sit for him. Smith replied, 'Friend, is thy servant a dog?'

Sydney Smith[20] was perhaps a reluctant Promethean. He shared their passion for reform and, despite impeccable Anglican credentials, argued persuasively for Catholic emancipation from the time of his editorship of the *Edinburgh Review* right up until the 1829 Bill which finally gave dissenters more or less equal rights. He hated war and the Game laws. He admired railways. He believed in a universal and liberal education and was almost as popular a lecturer at the Royal Institution as his friend Humphry Davy. He was, when in London, an essential ornament at that liberal palace Holland House and, indeed, at any table graced by the presence of his appreciatively large appetite.

Honesty, sometimes crushing honesty, and his politics excluded Smith from the highest offices of the Church, sometimes to his own regret. He did not identify himself with the Greek Titan of self-liberation but he was unafraid of the truth. Smith's analysis of the country's ills in the aftermath of Peterloo, sensibly confined to a private letter, would have served perfectly as a hustings speech for Shelley's unbound Prometheus:

> What I want to see the State do is to lessen in these sad times
> some of their numerous enemies. Why not do something for the
> Catholics and scratch them off the list? Then come the Protestant
> Dissenters. Then of measures, – a mitigation of the game-laws –
> commutation of tithes – granting to such towns as Birmingham
> and Manchester the seats in Parliament taken from the rottenness
> of Cornwall – revision of the Penal Code – sale of the Crown lands
> – sacrifice of the Droits of Admiralty against a new war; – anything
> that would show the Government to the people in some other
> attitude than that of taxing, punishing, and restraining.[21]

Other members of the Martins' circle, being more discreet about their politics, were welcome everywhere. A frequent visitor was young Charles Wheatstone, the friend shared by John with Michael Faraday. Whilst he was still unknown, and although terribly shy, Wheatstone could easily be induced to install some mechanical trick or other at the Martins' to impress all-comers. The artist and publisher Henry Viztely later recalled how Wheatstone had made a doll dance upon the grand piano, apparently independent of all mechanical aid. When pressed to reveal his secret Wheatstone replied, perhaps with a wink at Martin, that it was done by lightning, to which the assembly responded with hoots of laughter.[22]

When he was not walking, painting, or envisaging various engineering schemes, John Martin spent some of his happiest hours being read to by the eclectic Susan (according to Leopold, scarcely a single author of the day was unknown to them) or by exercising in the family sport of swordplay. He regularly paid a fencing-master to give him lessons in his painting room. Unlike his brothers, John also developed a (for his friends) disturbing passion for the javelin, which he used to propel with great energy against hand-painted targets on trees in what is now The Regent's Park.

In 1820 Martin made the acquaintance of two professionally useful brothers who, having been christened Allen, now called themselves respectively John Sobieski and Charles Edward Stuart and claimed descent from the royal Stuart line. In anticipation of a royal visit to Edinburgh by the Prince Regent, and tapping into a revival of Scots nationalism, these two amiable scoundrels forged a document called the *Vestiarium Scoticum* which they claimed to be an ancient depiction of the original clan tartans (in fact the clan tartan, like the pleated kilt, was a comparatively recent invention of the textile industry).[23] Hearing of their supposed expertise, Martin consulted them for his major canvas of 1820, *Macbeth*, now in the National Gallery of Scotland. Exhibited at the British Institution, the scene of the previous year's triumph with *Babylon*, it depicts Macbeth's return from the Highlands after his defeat of Macdonald, and his meeting with the Weird Sisters on the blasted heath. The subject suited Martin's treatment perfectly, its strong sense of windswept grandeur and the defiant figure of Macbeth no doubt a nod to current radical sympathies and to Martin's self-image.

A decade later, on his last visit to London, Sir Walter Scott, the ultra-Highland lowlander, called in at Martin's studio to see the work, which had never sold.[24] According to Leopold Martin, Scott expressed his deep regret at not being able to purchase the picture and John, by this time heavily in debt, expressed equal disappointment at not being able to give it to him.[25]

Regardless of the colour of John Martin's many friends, none were ever in doubt that he took his politics deadly seriously. Whatever interpretation may have been put on his pictures, his theology was Deist (the closest one could come to expressing atheism in public), his canon rationalism. There are strong echoes of Shelley's Prometheus in his beliefs, echoing the debt they both owed to William Godwin. According to Mary Pendered, John told his friend Ralph Thomas that: 'Man is endowed with a power of reason and everything is placed before him to reason by and with, as facts and arguments. Should veneration of his early education make him fear to approach his beliefs or religious tenets? Why should man use one logic for religion and a different kind for general affairs?'[26]

The beginning of the year 1820 marked the end of an epoch in British history. George III's 60-year reign ended with his death in January. The Prince Regent acceded to the throne ill (despite 80 ounces of blood having been let by his doctors, he somehow survived) unloved by his people, despised by his queen, heirless and in debt.

Before he could even begin the proceedings which he hoped would end in a divorce prior to his coronation; before, even, Parliament could be formally dissolved by custom, an attempt by a group of conspirators to assassinate the entire British cabinet rocked the political establishment. The so-called Cato Street conspiracy was partly the inspiration of a follower of Thomas Spence, Newcastle's land-sharing radical of the 1770s, called Arthur Thistlewood, and partly the idea of a government agent-provocateur keen to goad him and his fellow Spenceans into committing a rash act. Their plan, to murder the British Cabinet at dinner, was melodramatic enough. It was designed as the first act in a plot of national rebellion, timed perfectly to exploit the extraordinary state of unrest in the country. But only 27 men could be persuaded to join the plotters in their scheme to replicate the first days of the French Revolution. There were no powerful backers, few links to the Patriotic Unions of the working classes or the Hampden clubs of the Radical intellectuals. It was an isolated expression of zealotry. Nevertheless, the fight between the police and the plotters in their Cato Street loft, Thistlewood's killing of one of the officers with a sword, and the extent of their ambitions as revealed to the public at their trial in March, made for shocking headlines.

The government felt vindicated in its recent repressive legislation; the Opposition in their argument that reform must come soon to prevent outright revolution. The Cato Street trial, at which five were sentenced to be executed for high treason and the others transported for life, was immediately followed that spring by the trial and imprisonment of Henry Hunt, Samuel Bamford and others for their part in the Peterloo Massacre of the previous summer. The trial having been conducted at York, the guilty defendants were bailed to appear at the King's Bench in London for sentencing. Hunt arrived there basking in a tumult of public support. *The Times* reported that the crowds which welcomed him numbered three hundred thousand.[27] The sentences: two and a half years' imprisonment for Hunt, a year for the others, reflected both the government's determination to suppress dissent and its fear of creating martyrs by transportation or execution.

The nation had barely regained its breath before news arrived of the sensational return of Queen Caroline from her six-year sojourn on the Continent, during which she was widely supposed to have behaved in a very scandalous fashion. Her determination to return in the face of threats from both her husband and his government was in large part due to the machinations of her self-appointed counsel, Henry Brougham. Brougham had long seen in the cause of the injured Princess Caroline a chance to bring down Lord Liverpool's Tory administration and to promote his own career, via the silk, to become Lord Chancellor and then Prime Minister. His self-serving handling of Caroline's cause was a game played for high stakes, and not just his own. The queen's adultery, if proven, was treasonable. If she were found guilty, the Radicals were poised to ride a wave of popular anger and bring down not just the government, but the monarchy.

Brougham had withheld Prime Minister Lord Liverpool's offer to the princess in 1819 which might have kept her in the warm arms of her lover Pergami in Italy. In doing so he aimed to increase the pressure on the government while he assured the queen that he was acting in her best interests. Her snubbing by the papal court and the British Continental aristocracy on the death of George III (although the Davys, among others, visited her in Rome) determined her now to seek revenge by spoiling the king's coronation party. Brougham's desperate attempts to prevent Caroline from crossing the Channel at the beginning of June 1820 failed when she learned of his earlier duplicity. When she reached London on June 6th there were riots; those who failed to illuminate their windows in her cause had their windows broken. A bitter campaign was conducted in the press, in cartoons, in leaflets and on street placards in anticipation of the king's and the government's response. The weapon of public opinion, what Robert Peel called 'that great compound of folly, weakness, prejudice, wrong feeling, right feeling, obstinacy and newspaper paragraphs'[28] was wielded with increasing recklessness on both sides. In the middle of June one of London's Guards' regiments mutinied and the Duke of Wellington had to act quickly to prevent this spreading.

In July 1820 a Bill of Pains and Penalties to dissolve the marriage, expel Caroline from the country and deprive her of her titles was introduced by a reluctant government as a means of avoiding a judicial trial, although that was exactly how its passage was seen. It was clear to all that Caroline had in fact committed adultery just as her husband had, but Brougham defended her with consummate brilliance, managing to portray the government as vindictive, the pursuit of witnesses from Italy as corrupt, their testimony as perjury. The second reading of the bill was passed by an uncomfortable 28 votes; the third by just 9. By the middle of November the bill was dropped.

For five nights the cities of Britain were illuminated in celebration. The perverse outcome was that Caroline's public victory dissipated the mob's anger against Crown and government. Enough mud had been thrown to ensure that a nasty stench clung to Caroline and her largely foreign friends and staff. She threw away her moral advantage by accepting a financial settlement from the government. Liverpool's administration did not fall because the Whigs could not unite long enough to provide an alternative. The Radicals failed to capitalize on the queen as an admittedly double-edged political asset. Henry Brougham had confirmed his reputation as at once the most brilliant forensic advocate of his generation and an unprincipled opportunist. Whatever his future achievements, and they were many, he was never politically trusted again.

In July 1821, having been very publicly and humiliatingly turned away from the door of Westminster Abbey at her husband's coronation, the uncrowned Queen of England fell into a rapid decline, and by August was dead. Hers was not the only notable death that summer. Napoleon, too, had died, in his case still in exile, chained Prometheus-like to the rock of St Helena in the South Atlantic Ocean. His personality, ambition and politics had defined not just his own generation, but the future of European geopolitics. He may not have fulfilled the Godwinian vision of perfectibility but he had shown that the most audacious act of theft was within human grasp.

During the furore over the bill against the queen, John Martin had fallen out with his friend, the painter C. R. Leslie. According to Mary Pendered's account of Ralph Thomas's diary they were at a concert together, and when the National Anthem was played at the end Leslie cheered while Martin, as a partisan for the queen's cause, hissed. Leslie was so upset that he left. At that time he was hoping to be elected as a Royal Academician and Martin's behaviour, which soon became widely known, compromised his candidacy. They were never afterwards so close as they had been.

Leslie, as it happened, played an inadvertent part in John's great triumph during the year of the coronation. Some years previously he had introduced Martin to his friend Washington Allston, a fellow American artist. Allston warmly praised *Sadak*, and in turn John found a young man whom he came to admire greatly. They discussed Allston's conception for the subject of *Belshazzar's Feast* which he was then painting (and which, it having become a sort of *idée fixe*, he never completed). Martin, who in argument could quickly form an adamantine opinion, gave a quite different view of what he thought the painting should look like. Allston told him there was a poem written by a

T. S. Hughes which tallied exactly with Martin's idea, and that he should read it. Having done so, Martin determined to paint the subject himself. Leslie was among many friends who heard of John's intention and tried to dissuade him. As usual John ignored them all. He said to Leslie, 'I mean to paint it, and the picture shall cause more noise than any picture did before.' Then he added, as if realizing that this sounded a little pompous, 'Only don't tell anyone I said so.'[29]

Through 1818 and 1819 Martin worked on *Belshazzar's Feast*. There was much speculation among artists, scientists and mathematicians who heard of this grand experiment in historical painting. The curious came to Martin's studio on Allsop Terrace to watch progress for themselves. Mrs Siddons, the most famous actress of her day, came to admire it and kindly posed on the spot for Martin to get the attitude of the prophet Daniel just right. According to Leopold Martin, despite her advanced years she was still a striking woman: 'in spite of a very evident moustache, small traces yet remained of her former, queen-like, perfect beauty'.

Odd fragments of sketches for the picture's architectural elements survive in the Sezincote sketchbook and it is obvious in retrospect that the *Fall of Babylon*, successful as it was, constituted a sort of dry run for *Belshazzar's Feast*. The biblical story of hubris, folly and punishment was well known to the British public from the Book of Daniel. King Belshazzar of Babylon possessed great treasures which his father King Nebuchadnezzar had looted from the Temple of Jerusalem, and held a great feast for his court at which thousands drunk wine from stolen vessels of gold and silver. During the feast a mysterious hand appeared and wrote in magical letters on the wall the words '*Mene Mene Tekel Upharsin*'. The Jew Daniel was brought before the king to explain the meaning of these strange words. Daniel told him that the meaning was 'God hath numbered thy kingdom and brought it to and end; thou hast been weighed in the balance and found wanting; and thy kingdom has been divided among the Medes and the Persians.' The next day Belshazzar's kingdom was overrun by the Persians.

By the end of 1820 Martin's conception of the feast was ready to be sent to the British Institution. Martin's critics would have had in mind Rembrandt's famous rendition of this scene of 1635, an intimate study of King Belshazzar's horror of realization. Martin's public are more likely to have had in mind a caricature of James Gillray's which parodied Napoleon Bonaparte in 1803 as the tyrant whose time was up, gorging on his ill-gotten spoils in denial of the 'writing on the wall'. The unveiling of this new, eagerly anticipated work in February 1821 could not have been timed more poignantly. No monarch ever came to the British throne more unloved. This was a man whose coronation costume alone cost more than twenty-four thousand pounds, the salaries of a score of admirals. The splendour of his palaces, the magnificence of his feasts, his dissoluteness, adultery and treatment of his queen could not have been more like those of an Old Testament tyrant. To top it all, the new king's own Coronation Feast cost the obscene sum of £243,000. Not a single person could have been in doubt of the seditious, if opportune, nature of the painting which not only predicted George IV's downfall, but seemed to demand it.

John Martin's *Belshazzar* was more than mere sedition. It was the mature conception of a master of architectural perspective and dramatic lighting, of historical narrative and visual impact. The area around the painting at the Institution had to be railed off, such was the press of people eager to get close to it. The exhibition was extended by three weeks and the directors of the Institution had no hesitation in awarding *Belshazzar* the annual first premium of two hundred guineas. The *Magazine of Fine Arts* called it the 'most dazzling and extraordinary work'. The art critic of the *European Magazine* hailed it as 'a poetical and sublime conception in the grandest style of art'. The artist David Wilkie wrote:

> His picture is a phenomenon. All that he has been attempting in his
> former pictures is here brought to its maturity; and although weak
> in all those points in which he can be compared with other artists,
> he is eminently strong in what no other artist has attempted. Bels-
> hazzar's Feast is the subject; and in treating it, his great elements
> seem to be the geometrical properties of space, magnitude and

number, in the use of which he may be said to be boundless. The great merit of the picture, however, is perhaps in the contrivance and disposition of the architecture, which is full of imagination.[30]

Despite an offer of 800 guineas for the picture from the Duke of Buckingham (who, like the Duke of Sussex who admired it, seems to have been blissfully unaware of its seditious intent), Martin sold *Belshazzar's Feast* to his former employer William Collins in slightly mysterious circumstances for a sum said to be less than that offered by the duke. Collins did very well out of the deal: another 5,000 people came to see the painting at his premises on the Strand where he also sold a sixpenny pamphlet, written by Martin, describing the chief features of the picture. Collins also persuaded Martin to paint a second version of the picture on glass, which he also exhibited on the Strand, inserted in a wall and back-lit for even more sensational effect.

As with all John Martin's works *Belshazzar's Feast* was criticized for his diminution of the human figures, for what John Ruskin called its 'foxy hues' and its scandalous elevation of architecture over landscape. It was, above all, a picture for popular public consumption. In every sense it has stood the test of time less well than Constable's *Haywain* of the same year and Turner's sea-and-harbour works (*Shields, on the River Tyne* of 1823, for example) of about the same time. It is a matter of some luck that the *Feast* has survived at all, for it suffered a superbly Promethean accident the year after John Martin's death: in November 1854 it was being transported by carriage to Liverpool for an exhibition, when the vehicle was struck by a locomotive at a railway crossing.[31]

Comparison with Constable's work is of more than passing interest. The two men knew each other well. John and Leopold walked over the Heath to visit him on their rambles, always making sure to dress for the wet as Constable was usually to be found splashing about *en plein air* like his Impressionist inheritors. It is striking that Constable's finest work is to be found in his oil sketches; the static, polished versions of subjects like *The Haywain* have much less vigour and movement in them. It is as if Constable felt he ought to tame his natural impulses for the punters. In complete contrast Martin's oil sketches, such as that for Joshua, are mere planning exercises for the fully worked-up visual extravaganzas that he unveiled before his public like grandiose architectural narratives. Both men employed the sky to great effect; both believed that they were applying principles of science to fine art. One, neglected in his own lifetime, has achieved almost the status of a secular English saint. The other, hugely famous in his day, is almost forgotten.

There was no question that John Martin could be prosecuted for *Belshazzar*, even though the king, when Prince Regent, had urged the government to suppress the works of caricaturists like Gillray and Cruikshank for their satirical portrayals of him. Fine art, it seems, was above such imputations. Two Prometheans were, however, prosecuted and imprisoned in 1821. John Hunt, now living in semi-retirement in Somerset, was convicted of seditious libel in the same month that *Belshazzar* was exhibited. He had written a withering and unsubtle piece in *The Examiner* the previous July under his pseudonym Charles Fitzpaine (a nod to Thomas Paine) in which he described MPs as: 'Venal boroughmongers, grasping placemen, greedy adventurers and aspiring title-hunters ... a body, in short, containing a far greater pro-portion of Public Criminals than Public Guardians.'[32] He was tried and convicted in a single day and once more found himself incarcerated in Cold Bath Fields prison, protesting that his trial was intended to put down the demand for reform but would fail.

Marc Brunel also found himself in prison in 1821, although he was far too much of a loyalist to commit anything so outrageous as sedition. He was guilty of debt. After the failure of his army boot man-ufactory and the expense of rebuilding the Battersea sawmills, Brunel had busied himself with various engineering schemes. In 1816 he suc-cessfully installed a double-acting steam engine on the Margate packet ship *Regent* and at the same time dreamed of a great future for steam navigation: 'What a sight for the Chinese, by and by!' he wrote.[33] In 1817 he visited Paris for the first time in 20 years, commissioned to install steam pumps to supply the French capital with fresh water. His plans were confounded by the opposition of the city's *porteurs d'eau*, as militant a group of workers as the Tyne's keelmen. He had since then advised the Russian Tsar on the construction of a road across the River Neva and designed a suspension bridge for it with an 800-foot arch, not commissioned because the Tsar was supposedly too poor.

While his friends, including Lord Spencer and the Duke of Wellington, discussed how to help pay off his debts, Marc's wife Sophia came to stay with him at the King's Bench prison in Southwark. Eventually the government stumped up the funds to pay Brunel's creditors after he threatened to take his talents off to Russia permanently. Admiral Codrington, who had been deputed to bring him the news, remembered that: 'On visiting the prison, the deputation was ushered into a small room, in one corner of which was Brunel at a table littered with papers covered with mathematical calculations, while, seated on a trestle table in the opposite corner of the room, sat his wife, mending his stockings.'[34]

John Martin, it seems, represented a minority in profiting from his work. The ageing but inspirational William Godwin, living on Skinner Street in Clerkenwell, managed to fend off his debtors for just one more year. He had been subsidized by a number of patrons in the last decade as his various publishing ventures failed to meet his financial needs. Shelley had made himself broke with any number of gifts and loans to his mentor. His death by drowning in 1822 left Mary carrying the burden of support for her father, a challenge which she accepted insofar as she was able. She even sent him the manuscript of her second novel *Valperga* (1823) for him to publish and profit from.

Michael Faraday remained at the Royal Institution, growing in popularity as a brilliant lecturer, sometimes helping his mentor Sir Humphry with experiments, and spending more and more time exploring the mysteries of gases. A handsome, tousle-haired 30-year-old with uncommon abilities and a mind of great precision, he had by now formulated his own version of Promethean perfectibility in a lecture to the City Philosophical Society in which he warned his fellow scientists of the follies of hubris:

Nothing is more difficult and requires more care than philosophical deduction, nor is there anything more adverse to its accuracy than fixity of opinion. The man who is certain he is right is almost sure to be wrong; and he has the additional misfortune of inevitably remaining so. All our theories are fixed upon uncertain data, and all of them want alteration and support. Ever since the world began opinion has changed with the progress of things, and it is something more than absurd to suppose that we can have a certain claim to perfection; or that we are in possession of the acme of intellectuality which has, or can result in, human thought.[35]

This strong sense of realism and humility, of his place in a line of philosophers who sat on the shoulders of giants,[36] did not prevent Faraday from aspiring to the discovery and understanding of new phenomena. Far from it. But in 1821, recently married to fellow Sandemanian Sarah Barnard and now on the brink of making some of his most important discoveries, he committed a mistake of fatal naïveté which nearly brought about his premature downfall.

In July 1820 the Danish philosopher Hans Christian Oersted announced to the international scientific community his discovery of the phenomenon of electromagnetism, after he had induced a compass needle to be deflected by a weak electric current. This exciting new relationship between these two energies immediately struck chords with those who, like Oersted, followed Emmanuel Kant's view of the unity of natural forces. Philosophers across Europe rushed to their laboratories to repeat his experiments, suspecting rightly that here stood ajar a door to a vast new area of research, and at the same time wondering why none of them had tried it before.

In Paris André-Marie Ampère was shocked into reformulating his theories on the nature of the electric fluid to fit into a comprehensive model of the nature of matter. In England Humphry Davy, who many years before had been introduced to Kant's philosophy by his friend Coleridge, burst into Faraday's laboratory at the Royal Institution to tell him the news. Both men at first misunderstood the implications of Oersted's conclusion, that electromagnetism was a radiating force. Faraday was too busy with experiments on steel alloys and on chlorine and, perhaps, his forthcoming marriage, to pay much attention. Davy continued his work on this new phenomenon with Dr William Hyde Wollaston, a highly respected experimenter who correctly predicted that two wires carrying a current would attract each other if the currents flowed in the same direction and repel each other if they flowed in opposite directions. Wollaston also saw that a wire would rotate on its own axis when subject to a magnetic force.

Faraday himself only began serious experimentation on the phenomenon in the summer of 1821 when his friend Sir Richard Phillips asked him to contribute a paper on electromagnetism to the *Annals of Philosophy*. Overwhelmed by the many seeming contradictions in the theories of his colleagues at home and abroad, he proceeded with the caution inherent in his scientific philosophy, checking everything by experiment. His instinct told him that although the arguments in favour of electric current as a fluid implied by Ampère's, Davy's and Wollaston's work were strong, there was a small possibility that electricity was a 'state', and this he made clear in his paper.[37] What he saw as Oersted's and Ampère's belief in two distinct types of electric fluid flowing in opposite spiral directions he dismissed.

Having published the paper anonymously Faraday continued with his own experiments in September 1821 and very quickly found something quite new and exciting. Wollaston had predicted the rotation of a wire around its own axis; Faraday proved that a wire would rotate around a magnetic pole, and in so doing discovered the effect that led him, ten years later, to a grand theory of electromagnetic induction. Magnetism induced rotation, and its effects varied precisely with the electrical conductivity of materials. Faraday went to Wollaston to tell him the news and discuss its implications, but Wollaston was away. Wary of taking his colleague's name in vain, Faraday published his news in the *Quarterly Journal of Science* in October, omitting Wollaston's name entirely. It was a perfectly reasonable course of action but a disastrous mistake for a young man working among exalted company. He was very soon openly accused of having stolen Wollaston's ideas without crediting him, though in fact what he had discovered went far beyond Wollaston's observations. Both Wollaston and Davy refused to support him during the furore that followed. Opprobrium rained down upon this upstart. Faraday was mortified, his chance to step out of Davy's shadow apparently wrecked by his unwitting Promethean folly.

The Prometheans were playing for high stakes. Only the painters Turner and Martin seemed to be riding the wave of public opinion. John Hunt and Marc Brunel were in prison. Faraday's nascent career lay in ruins. Shelley, in self-imposed exile, was unpublishable. Brougham's ambitions had been thwarted by his own disingenuousness and naked ambition. Charles Babbage seemed too brilliantly cerebral to achieve anything. These men's fortunes recovered in their various ways. There was, however, one Promethean whose career was so fatally compromised in 1821 that he was forced to flee England, never to return.

Friedrich Accum had been doing rather well since his pioneering work on gaslight. His chemistry sets and laboratory appliances sold well and he had taught a succession of gifted pupils who were now established professors at Harvard and Yale in the United States. In 1817 he published *Chemical Amusement* containing experiments for interested amateurs to try at home using his own portable chemistry sets which ranged from five to twenty guineas in price. But his greatest contribution to applied science came with the 1820 publication of his *Treatise on Adulterations of Food and Culinary Poisons*, published by Rudolf Ackermann and beautifully illustrated. Accum, who had made a painstaking study of the means by which unscrupulous manufacturers substituted cheap and often dangerous chemicals for proper ingredients, was asking for trouble, and he knew it: 'However invidious the office may appear, and however painful the duty may be, of exposing the names of individuals who have been convicted of adulterating food, yet it was necessary.'[38] His exposure of the shady world of brewers (who, when they were not diluting their beer, used a variety of poisonous substances to give it flavour and a better head), tea-suppliers (who recycled used tea leaves by drying them on copper plates to give a green patina), purveyors of pepper who used red lead to enhance its colour, and many others, resulted in a number of prosecutions. The book went to two editions in its first year and was also published in Germany and America. Accum began to receive anonymous threatening letters. Within months of the appearance of the second edition the Royal Institution received a tip-off that Accum had been 'mutilating' books in their collection, of which he was a former librarian. Pages from some of their books were apparently found during a search of his apartment, though they proved to be only waste paper. He was acquitted on a charge of robbery. The Institution then undertook its own proceedings against him and he was charged once more. Whether or not Accum had in fact stolen or mutilated books was never established; the whole episode seems bizarre. In a fit of depression exacerbated by the public ignominy of the proceedings against him he failed to appear at his trial in April 1821 and a few months later fled to Germany in disgrace. He had paid a high price for tilting at giants.

CHAPTER TEN

Paradise Lost

There went a fame in Heav'n that he ere long
Intended to create, and therein plant
A generation, whom his choice regard
Should favour equal to the Sons of Heaven.[1]

John Milton (1608–74) the puritan poet and republican propagandist who survived the English Civil War [2] and Interregnum before being arrested and nearly executed in 1659, was almost 60 when he published his masterpiece *Paradise Lost*. Blind from glaucoma, he dictated the entire epic poem at the cottage in Buckinghamshire where he had withdrawn to escape the plague raging in London in 1665. He was lucky to miss the Great Fire that destroyed most of the capital the following year. Even so, his was an apocalyptic world-turned-upside-down full of redolence for the Private Prometheans.

Paradise Lost, published in ten books in 1671, relates the fortunes of Adam and Eve, their creation and fall from grace, their expulsion from Paradise and the sacrifice of God's only son to redeem their guilt. But it is the figure of the fallen angel Satan, Adam's bane, who emerges as the poem's emotional protagonist in the war he wages against God. The Prometheans were torn between their veneration of the Greek Titan of insolent liberation and the attractions of the Satan figure. As Shelley had argued in his preface to *Prometheus Unbound*, the Greek hero was a purer, more satisfyingly redemptive figure than Satan, while the latter offered a more deliciously subversive and ambivalent anti-hero with whom to experiment poetically and philosophically.

There is no doubt that Milton's own career and the lyrical strength of his poetry made *Paradise Lost* an essential component of the intelligent (or aspiring) reader's library in this period. Illustrations by Fuseli, Blake and others of the Gothic Revival ensured its popular appeal. But, as Shelley had realized, neither Satan nor Adam were entirely satisfying figures and Prometheus's fire and chains were irresistible metaphors in an age of revolution. The collective output of a generation of fire-bringers suggests a conflation of Prometheus, Satan, Mephistopheles, Adam and Christ into a single sacrificial, insolent, liberating and redemptive figure. They saw themselves as the generation whom, in Milton's own words, God's 'choice regard should favour equal to the Sons of heaven'.

Despite many of them having been born into poor families with no independent wealth or status, the Prometheans had begun to achieve social and financial liberation in the first two decades of the nineteenth century. The fruits of the Industrial Revolution were theirs to pick: new wealth and a high demand for innovation; a reduction in childhood mortality and consequent surplus of aspiring youths of both sexes; dramatic changes in travel and communication; a physical and intellectual world seemingly growing larger as new horizons were explored and conquered. This fertile ground was seeded by the technical and personal imperatives of war, which has always successfully exploited talent. It is entirely fitting that the Prometheans, epitomizing this generation, should feel favoured by God's choice regard.

For much of the nineteenth century popular versions of *Paradise Lost* and the Bible illustrated by John Martin ensured his enduring presence in thousands of homes of every class. The combination of an artist of sufficient imagination and tireless energy with a technological printing innovation which enabled ordinary families to possess such marvels is a Promethean tale par excellence.

Among John Martin's associates in the early 1820s was Thomas Lupton, a rather handsome man two years his junior. Born in

Clerkenwell, the son of a goldsmith, he was apprenticed to the engraver George Clint in 1805, the year before Martin arrived in London. He exhibited a number of portraits at the Royal Academy and made his reputation with engravings of several subjects in J. M. W. Turner's *Liber Studiorum* series of prints. The two men became firm friends. In 1822 Lupton made the first successful steel mezzotint engraving, launching a medium that was to popularize nineteenth-century art because for the first time large numbers of prints could be taken from a single plate before its surface wore away.

Before Lupton's pioneering process, prints of illustrations had been taken using woodcuts, wood engravings, copper mezzotint and aquatint. By the end of the eighteenth century wood engraving had reached a technical and aesthetic peak in the work of John Martin's fellow Northumbrian Thomas Bewick (1753–1828). Widely known today for his vignettes of countryside and naval scenes, his greatest work was executed in natural histories of birds and quadrupeds. The fineness of his execution owed much to the technique of carving boxwood across the end grain, in contrast to woodcuts which were carved with the grain. The durable, dense fibres of box allowed large numbers of prints to be taken and made the technique quick to execute. But the size and scope of such engravings was limited by the material. As public demand for landscape and especially architectural designs grew and as the physical possession of beautiful books became fashionable among the middle classes, engravers responded with a range of new techniques.

Etching and engraving on copper plate had been established techniques before the eighteenth century began. Both involved laborious processes. In an etching the plate was coated with waxes and resins, the design scraped through them and then etched with acid. In a mezzotint the plate was prepared with a rocker to finely indent its surface. When inked this surface produced a matt black print. The engraver worked this surface with a variety of scrapers to produce a range of tones up to pure white. Artists were naturally concerned that the engraver should do their work justice, and in response to technical or contractual failures many learned the techniques themselves.

A more sophisticated method of mezzotint, introduced in the middle of the eighteenth century, involved stippling the copper plate through an etching ground and then scraping a design through the stippling to produce distinctive soft tones that were ideal for landscapes and skies. Then, towards the end of the century, a new process called aquatint was invented in France by Jean Baptiste le Prince. Le Prince used a porous coating through which acid burned first the lightest parts and then, after a coating of protective varnish, successively darker layers. Aquatint was widely adopted for technical prints (it was a favourite of Rudolf Ackermann, for example, in his various treatises) while mezzotint was used in reproductions of portraits and paintings.[3]

When the market for monochrome or hand-coloured book illustrations and popular prints expanded at the turn of the nineteenth century, the limitations of the copper plate forced engravers to experiment with new media. Copper's softness, which gave mezzotints such attractive tonal subtlety and warmth, limited the number of prints which could be taken from a plate before the surface wore away. Mass-market book production and copper mezzotint were mutually exclusive.

The first successful print process using a harder steel plate was used in America in 1810 to produce a new banknote.[4] A steel surface could produce upwards of a thousand impressions before significantly losing quality. However, because the surface of the metal was harder than copper it was also much more difficult to prepare and scrape and failed to hold ink with anything like the saturation possible with copper.

In England the first to experiment with steel mezzotint was William Say in about 1820; but he gave up. Two years later Thomas Lupton successfully reproduced George Clint's portrait of the actor John Shepherd Munden, having employed a combination of etching and scraping and developed a more adhesive ink. It won him the Isis medal from the Society of Arts and much interest from his painter and publisher friends.

The deficiency of the steel mezzotint was immediately obvious. The unyielding surface of the steel plate translated into a thinness, a coldness, in the print. The problem was not fully solved until near the end of the century when an engraved copper sheet was plated with steel. In the meantime, publishers of illustrated books were not slow to commission designs for ambitious mass-market projects, mitigating the deficiencies of the new process by experimenting with tools and inks, and careful choice of subject.

John Martin was particularly receptive to Lupton's new process. He had, to begin with, a love of engineering and invention, much like Turner. He must have seen that the coldness of the steel mezzotint would actually suit his linear, architectural designs. He had also been frustrated for some time over an engraving of *Joshua* which he had commissioned from Charles Turner (who had executed the first 20 illustrations in the *Liber Studiorum* of his unrelated namesake). By 1822 the promised plate had still not been completed.

In any case, after the reception of his masterwork the previous year John was looking for a new challenge. He decided to attempt an engraving of *Belshazzar's Feast* himself and was busy learning the techniques of copper mezzotint when Lupton's invention appeared. He immediately scrapped his copper plate and persuaded Lupton to teach him the new steel process.[5] It was while he was experimenting with rockers, burins, scrapers and any number of combinations of inks, that the publisher Septimus Prowett commissioned him to design and engrave 48 illustrations for a new high-quality edition of *Paradise Lost*.

The commission, probably offered at the beginning of 1823 and worth an initial £2,000, came during a hiatus in Martin's career. He had been aware even before its exhibition that *Belshazzar's Feast* would be his defining work in the public mind, and he was right. How could he follow it? His first plan was to mount a solo exhibition at London's largest private gallery, the Egyptian Hall in Piccadilly in 1822. Here he was able to show all his major works (*Belshazzar's Feast* apart). The centrepiece was a new large-scale canvas, the *Destruction of Pompeii and Herculaneum*, which he had promised the Duke of Buckingham. It is hard to evaluate the conception or execution of the painting, since it was destroyed by fire in 1928; but while the exhibition cemented Martin's fame as the principal imaginative artist of his day, it seemed also to mark the passage of his immediate dramatic energies. Now he plunged himself into another apprenticeship, determined to master steel engraving as he had mastered china painting, architectural perspective and narrative historical drama.

J. M. W. Turner would have been equally interested in the potential of the steel mezzotint (he later employed Lupton for his *Ports of England* and *Rivers of England* series) but he was absent from London during the latter part of 1822. In August he made his way by ship along the North Sea coast to Edinburgh to be present during the visit of George IV to Scotland. After the constitutional débâcles of Caroline's 'trial' and his queenless coronation, King George undertook an exhausting public relations campaign during which he succeeded in reversing his popular fortunes. In Ireland, in Hanover and now in Scotland he was greeted by huge crowds and an outpouring of royalist sentiment. The Scottish trip was masterminded by Sir Walter Scott who, in the process of almost single-handedly inventing the Highland myth of picturesque tartaned clans, somehow managed to associate the new king with a revived sense of Scottish identity. It was a nice trick to pull off.

For Turner the exercise proved equally successful, less perhaps for George's appearances themselves than for the sketches of North Sea ports and harbours which the artist visited along the way. He spent days among the keelboats, colliers and staithes of Shields at the mouth of the River Tyne.[6] Here Turner compiled some of the most evocative and detailed surviving reportage of a region in the midst of breathless industrial and social change – shot-tower and all.

By coincidence 1822 was the year of the great Tyne keelmen's strike. Turner's *Hurries – Coal Boats Loading* with its smutty bustle and sense of intense activity (which other artist of the time was so interested in such 'low-life' subjects?) seems to capture the precise moment in the keelmen's history when new technology was about to deal the decisive blow to their ancient trade. Here are the new staithes, coal chutes and railway wagons loading coal directly into colliers. Turner's treatment of them is intensely interested without ever becoming partisan or sentimental: it is the finest sort of journalism.

Their ancient livelihood and privileges threatened by such innovations, the keelmen blockaded the Tyne upstream at Newcastle in October. Those collier-viewers whose operations still relied on keels because the upper waters of the river were too shallow for London-bound colliers to moor, responded swiftly. Among the foremost was William Hedley. There are few more telling scenes in industrial history than the contemporary print showing one of William Hedley's old locomotives, *Wylam Dilly*, shorn of its wheels and hastily adapted to drive an improvised steam paddle tug, hauling coal barges downstream beneath the Tyne Bridge. And so 1822 proved to be quite a year for steam shipping. For further south, in June, the *Aaron Manby* made the first iron-hulled steam-powered crossing of the English Channel.

Close to the other great northeast river, the Wear, another colliery engineer, George Stephenson, had just begun construction of the world's most famous coal-and-passenger railway between West Auckland, Darlington and Stockton-on-Tees in County Durham (the first public railway was already operating in Surrey, but it employed no locomotives). Almost overwhelmed by the constructional difficulties of such an audacious enterprise, Stephenson benefited from the keelmen's misfortune as hundreds of them flocked to work as his navvies (or navigators: the term derives from gangs of canal excavators active a century before).

Darlington, the principal wool market town of the northeast and a centre for Quaker entrepreneurs, was about to become the vibrant capital of the early railway industry. It was here that Philadelphia agent William Strickland came to report back to the United States on British industrial technology (Stephenson let him make engineering drawings of his locomotives).[7]

It was in Darlington, in 1822, that convicted lunatic Jonathan Martin chose to settle after his dramatic escape, celebrated in verse and engraving by his brother William, from an asylum in Gateshead. Having been convicted of threatening to shoot the Bishop of Oxford, Jonathan was originally incarcerated in a private asylum in West Auckland. Here he was trusted with work in the owner's fields but hated his night-time confinement in a small room with four other inmates. He was often cold and hungry and, when drunk, the owner Mr Smith was violent and abusive to him. At times Jonathan was confined alone and chained to his bed. Eventually intercessions with the magistrate by his friends resulted in a transfer to a much more orderly asylum at Gateshead, run by a sympathetic keeper called Nicholson.

Jonathan was able to earn a little money from small jobs and at some point made enough for a coach fare to be sent to his wife and son so that they might visit him. But his earlier ill-treatment, his religious mania and bouts of fasting made him weak and miserable. Nicholson allowed Jonathan to attend local chapel services if he promised not to fast, but when he sold the asylum to a Mr and Mrs Orton during the winter of 1818 a new and much less accommodating regime was ushered in. There was immediate conflict over the prayer meetings that Jonathan organized for his fellow inmates. After seeing that he showed few signs of insanity the Ortons gave him a degree of freedom which he abused, in the summer of 1820, by leaving the asylum and taking the keepers' keys with him.

First he walked 20 miles to Hexham to visit his relatives and then a much greater distance to Norton, with the intention of showing the magistrate there that he was not a lunatic but should be allowed to go free. The magistrate, not surprisingly, had him arrested by the constable. Returned to Gateshead after three days of liberty he was placed in irons and chained to his bed, where he soon became infected with lice. As further punishment his wife and son Richard, both now living near the asylum in Gateshead, were refused access to him. The next time he saw his wife, a year later, he found that she was dying from breast cancer.

Jonathan's conditions slowly improved as he made himself useful to his keepers by tending the needs of the other inmates and working in the garden; but they never again allowed him off his chains, which were a constant source of aggravation to him. One day, during the summer of 1821 when he was in a part of the garden which gave a fine view of the Tyne Valley towards the place of his birth, a voice from on high spoke to him:

> I had an assurance given me that I should soon travel the same road again to the astonishment and confusion of my enemies. This set me on contriving how it could be effected, and looking down to

> my fetters, which were the principal obstacles, I said, 'surely
> some means may be found to get them off'. Observing a small
> piece of freestone, I recollected that freestone would reduce iron
> by rubbing. I put it in my pocket, resolving to rub off the rivets …[8]

It took Jonathan three days of more or less incessant work to weaken his fetters to the point where he could break one of them against the fender of his fireplace. He had, some time previously, noticed that in the closet of the room he shared was a draught of fresh air. Now, he began to poke at the ceiling in the closet with a stick until he made an opening into what proved to be the eaves of the building. Climbing into the roof space, and giving thanks for his former career as a foretopman in the navy, Jonathan found that he could loosen enough tiles to create a hole through which he might crawl. Thus satisfied, he returned to his room, extracted a promise from his fellow inmate not to breathe a word, and swept the debris from his exertions into the closet.

Suspecting nothing, the keeper came for his evening visit, shaved the two men, then fastened the window shutters and locked them in for the night. Later, when the full moon rose, Jonathan made his way through the closet into the eaves and then on to the roof with a small bundle of clothes. Having thrown his bundle onto a half-cut haystack he managed with great difficulty to lower himself onto the 18-feet-high perimeter wall and ran along it for 40 yards until he came to the place where a shed had been built against it. From the roof of the shed he jumped onto a dunghill, and was free. 'I had now my bundle to seek, which was soon done; a mastiff in the yard heard my foot and barked, but I was soon out of reach, and having gone a small distance from the place of my late wretched domicile, I felt as if I could fly.'[9]

Jonathan had many adventures before settling in Darlington. He was hunted across Northumberland and fled first to Ayr, where he had friends, then to Glasgow and Edinburgh, in both of which towns he believed himself to have been recognized. At one time he thought of making his way to London where his brothers John and Richard would surely shelter him. But having visited his oldest brother William at Wallsend, he was persuaded to return to County Durham to be reunited with his son. At the beginning of 1822 he found work at a tannery in Darlington and for some years settled there. His colourful autobiography, the first edition of which was published in 1825, sold many thousands of copies and eventually went to six editions.

For a time in the early 1820s it must have seemed as if a generation's dreams had come to nothing. For all their efforts there had been no political reform; slavery still existed; Catholics had not been emancipated; Rotten Boroughs were still rotten; the labouring poor of London and the industrial north were still impoverished, ignorant, sickly and disenfranchised. Many bright lights had been, or were about to be, extinguished. Leigh Hunt had gone to Italy in 1822 to found a new liberal journal with Shelley and Byron, only to find Shelley's drowned body on the beach at Viareggio with a volume of Keats' poetry in his pocket. Keats was dead too, of tuberculosis. Byron was soon to leave for Greece on an adventure in the romantic cause of national liberty and would die at Missolonghi in 1824. William Blake, Henry Fuseli, Humphry Davy: all dead before the end of the decade. William Godwin, bankrupt, was a spent force. The energy that had driven the Prometheans through war, poverty and repression to the edge of revolution seemed now to have been dissipated. Even John Martin, who looked as if he might have run out of apocalypses to portray, had begun to dedicate works to the king whose name he had so recently hissed at. Turner was painting a highly theatrical version of the *Battle of Trafalgar* on the king's commission. Sedition had become unfashionable.

The critic William Hazlitt, summing up this generation in his 1825 collection of essays the *Spirit of the Age*, reflected a general sense of malaise, of apathy, of ennui: 'The present is an age of talkers, and not of doers; and the reason is, that the world is growing old. We are so far advanced in the Arts and Sciences, that we live in retrospect, and doat on past achievements ... What niche remains unoccupied? What path untried?' [10] Hazlitt was talking in particular of Samuel Taylor Coleridge, whose staggering intellect had promised so much and yet seemed to deliver so little; his early revolutionary zeal, like that of Wordsworth and Southey, lost with middle age. 'There is no subject on which he has not touched,' wrote Hazlitt, 'none on which he has settled.'[11]

Hazlitt may have been in tune with the prevailing mood in 1825, but he was to be proved wrong. The surviving firebringers were gathering their strength. The railway age, electricity, global communication, political reform, factory legislation, public education and health were to be the fruits of their labours, though the great weight of establishment inertia had still to be overcome.

For those with their ears close to the ground there were rumblings to be heard. Literally, in the case of the Brunels. The year before *The Spirit of the Age* was published they began the seemingly impossible engineering task of tunnelling beneath the Thames from Rotherhithe to Wapping. Richard Trevithick had already failed in the attempt. Marc Brunel, released from the debtors' prison and inspired by the astonishing labours of the timber-boring worm *teredo navalis* in the Chatham dockyards, conceived of a shield constructed of iron which would constitute the face of his workings and allow hewers to excavate in safety across the whole height and breadth of the tunnel simultaneously. Bricklayers would follow behind as the three sections of the shield were propelled slowly forward by a system of levers and jack-screws.

Having spent much time preparing drawings for this unique apparatus, Brunel turned once more to Henry Maudslay, who set about constructing the monstrous new machine. Maudslay's works at Lambeth meanwhile continued to produce machine tools of ever-greater precision. Perhaps more importantly, it had now begun to produce some of the great mechanical engineers of the nineteenth century. Joseph Clement, son of a Westmorland weaver, and Joseph Whitworth, the son of a schoolmaster from Stockport, had both served time with Maudslay and were now embarking on their own illustrious careers. Clement was to be associated, for good or ill, with the young mathematician and visionary engineer Charles Babbage, while Whitworth contemplated super-flat surfaces and identical screws and dreamt of an instrument able to measure tolerances of a hundred thousandth of an inch.

Meanwhile, on the Thames, boreholes were sunk across the river to determine the most suitable depth of strata to penetrate and in 1825 excavation began at Rotherhithe on the first of the two roadways under the supervision of Marc's son Isambard Kingdom, now 19 years old and showing much promise as an engineer. The tunnel, the first practical attempt to bore under any large body of water, was a project of great interest among Londoners who had in the last decade seen a rash of bridges springing up across the river at Vauxhall, Waterloo and Southwark, though none east of the city. At that time nearly four thousand passengers were making a daily crossing below London Bridge by ferry, so investors (including the technophile Duke of Wellington) doing their sums saw a good potential return on their capital. The tunnel was of particular interest to J. M. W. Turner, proprietor of a tavern called the Ship and Bladebone close to the proposed terminal of the tunnel on the north side of the river at Wapping.[12] It looked to him as if business would soon be picking up; but he had to wait a long time to see a return on his investment.

Turner was among an elite group who in 1824 founded a new club called the Athenaeum, intended to provide congenial surroundings and company for 'literary and scientific men and followers of fine arts'.[13] It was proposed by John Wilson Croker, the Secretary to the Admiralty, in a letter to Sir Humphry Davy after a meeting at publisher John Murray's offices on Albemarle Street, close to the Royal Institution. Davy was enthusiastic and the project went ahead. At the first meeting it was decided to confine membership to 500 men, generally drawn from the Royal Academy and Royal Society, although this limit was later expanded.

The Athenaeum's first, albeit temporary secretary, was Michael Faraday who, in spite of opposition from his former mentor Davy, had recovered his reputation and been elected a Fellow of the Royal Society that year. His growing independent status was cemented by the isolation of benzene and other liquefied gas products. Other founding members of the new club included Sir Walter Scott, Sir Thomas Lawrence, Turner, Lord Palmerston and Samuel Rogers, a poet-banker of unimpeachable taste, invitation to whose breakfast parties was a badge of social rank beyond price. John Martin had been introduced to many of the great men of state at one of these breakfasts after having been accosted by Rogers at a dinner held by Sir Robert and Lady Peel. Martin and Rogers seem to have spent their time discussing the defecating habits of the cows Rogers used to watch from his window in St James's Place.

The Duke of Wellington, Prince Leopold, the Duke of Sussex and Home Secretary Robert Peel lent the new club gravitas and political rank. A grand new building designed by Septimus Burton for the club's permanent home on Pall Mall helped establish the Athenaeum as one of the greatest of London clubs. A few years later, in 1830, it was decided that nine eminent persons could be elected annually under what became known as Rule II; after that it became a sort of Prometheans' drop-in centre. John Martin was elected in 1832; Sydney Smith, Charles Wheatstone, Robert Stephenson, John Buddle, John Stuart Mill, Charles Dickens, Charles Darwin and his erstwhile captain Robert Fitzroy, W. M. Thackeray and many others joined over the next few years.

Some of the early members of the Athenaeum were active in support of Henry Brougham's Society for the Diffusion of Useful Knowledge (SDUK), which he founded in 1826 to impart 'useful information to all classes of the community, particularly to such as are unable to avail themselves of experienced teachers, or may prefer learning by themselves'.[14] With close links to the provincial Mechanics' Institutes, the newly founded London University (in whose creation Brougham also played a prominent part) and publishers including Charles Knight, the SDUK, commissioned educative works like the *Penny Magazine* and *Penny Encyclopaedia* to popularize history, science and literature. Among other items the first issue of the *Penny Magazine* in 1832 carried articles on Poland and the history of beer, and a piece entitled 'Excellence not limited by station'. The Society at once attracted charges of being 'democratical'. Elitists like Coleridge warned that popularizing knowledge would end merely in its 'plebification'.[15] Thomas Love Peacock, satirical novelist and former friend of Shelley, declared open season on Brougham (whom he facetiously called the Learned Friend) and his plebifiers. His Reverend Dr Folliott, very recently mugged in the fields close to the fictional *Crotchet Castle*, intoned what must have been a widespread fear among the genteel classes: 'Robbery perhaps comes of poverty, but scientific principles of robbery come of education. I suppose the learned friend has written a sixpenny treatise on mechanics, and the rascals who robbed me have been reading it.'[16]

Unfortunately for the new aspiring middle classes, it was not just the starving poor who might learn the scientific principles of robbery. Septimus Prowett's generous 1824 commission for the *Paradise Lost* plates ought to have ensured financial stability for John Martin and his six surviving children. Learning and developing techniques for the steel mezzotint process and preparing such a large number of unique designs for printing absorbed much of John's energy. He had already built a painting room in the back garden at Allsop Terrace; now he constructed a printing studio, attached to the house by a long gallery supported on cast-iron pillars. The new studio was filled with flywheel and screw presses built by Bramah and Maudslay. There were glass and iron ink-grinders, French, Indian and English drawers for canvases, blankets, inks, whiting and leather shavings; outdoor cupboards for ashes and charcoal. In order to overcome the inherent coldness of the steel mezzotint process John experimented with a range of inks, oils and whiting, with dashes of burnt umber 'just to give a warm tint to the cold white'.[17]

It was around this time that John took on his only commercial pupil, a young Irishman who gloried in the name of John St John Long. The Martins took him in and John taught him for three months. The student hoped to paint a great picture on the state of Ireland but, although John was encouraging, he had little hope of its success. According to Leopold Martin, Long was 'without talent as a painter, but had great enthusiasm'.[18] Having failed as an artist Long removed himself to Harley Street and set up as a quack tubercular doctor, with disturbing consequences some years later.

As if these projects were not enough, John's old friend Charles Muss had just died and John now undertook to supervise the completion of his unfinished commissions for William Collins. He also painted a number of watercolours and exhibited another major canvas, the *Seventh Plague of Egypt*, at the newly established Society of British Artists (he was one of its founders). Despite mixed reviews the picture, now in the Museum of Fine Arts in Boston,[19] was sold for 500 guineas to the Earl of Durham. But John's hopes of enduring security were dashed in September when rumours began to spread that his and the Musses' bankers, Marsh Sibbald and Co., were about to suspend payment. He quickly drew a large cheque on them, but it was too late. The senior managing partner in the firm, Henry Fauntleroy, was found to have forged stocks and securities worth many tens of thousands of pounds in order to conceal the bank's difficulties over a series of bad debts. In November 1824 he was convicted of forgery after a trial lasting just five hours. He was publicly hanged, watched by a crowd of more than a hundred thousand people.

Fauntleroy's execution was the last for forgery. Home Secretary Robert Peel was already in the process of reforming the criminal justice system in England, reducing the number of capital offences from 160 in 1822 to about 60 a year later. His penal reforms, including the establishment of the first police force by 1829, continued until 1838, after which no one was executed for a crime other than murder or high treason. But they came too late for Fauntleroy and were of no comfort to John Martin, whose long financial decline now began.

Peel was not the only reforming minister in Lord Liverpool's more relaxed Tory administration of the 1820s. William Huskisson pushed through measures to reduce tariffs on foreign trade. George Canning openly advocated religious tolerance for Catholics. There was less progress on repeal of the loathed Corn Laws – understandably since the British legislature was still composed almost entirely of landed interests; but in 1824 the Combination Acts were repealed, giving workers a limited right to strike and to form trades unions.

One remarkable piece of legislation passed by Parliament in 1822, confirming a subtle shift towards a more humane vision of society, was MP Richard Martin's Act to Prevent the Cruel and Improper Treatment of Cattle. No relation of John Martin, 'Humanity Dick', as he was known, was derided by cattle breeders who resented outside interference in their business and by a labouring poor who might have wondered why cattle were being thus favoured by the government when they were not. The bill was an easy target for satirists, including the anonymous author of a curious novel which, but for its fame under another name, must have been forgotten. The *Rebellion of the Beasts* was published in 1825 by 'J and HL Hunt'.[20] The title page names the author only as 'a late Fellow of St John's College, Cambridge', but the novel has been widely attributed to Leigh Hunt who, if he wrote it, must have penned the manuscript in Italy and sent it to his brother without ever referring to it in a letter. It satirizes not only the revolutionary ardour of animals emboldened by their new rights; it also takes a swipe at monarchy in general and George IV in particular. Its similarities to George Orwell's *Animal Farm* of 1945 are more than passing (although no one has yet demonstrated that Orwell knew of it): the theme is pro-reform, the treatment cynical as the self-liberating beasts set up a republic with a bill of seven rights which is perverted when the Ass assumes the tyrannical characteristics of his human predecessors. Whoever the author, no one was prosecuted for it under the libel laws.

Not the least impressive work of John Martin's staggering output of the mid-1820s was another large (66 by 102 inches) oil painting, *The Deluge*, which he exhibited at the British Institution (having already fallen out with the Society of British Artists).[21] It has the same foxy hues, the trademark bolt of lightning, the diminished figures of his other apocalyptic works, another Old Testament warning to contemporary society of its follies and destiny. The work never sold and attracted little public attention. One Scottish critic was appalled by John's portrayal of miniature humans, 'the demented survivors o' the human race a' gathered together on the ledges o' rocks – there's thousands on thousands o' folk broken out of Bedlam and as mad'.[22] It was an unfortunate reference, as things turned out. But years later, when John was retouching it, as he often did with old paintings of his, it was an object of scientific interest from a highly eminent visitor. John was not at home when the gentleman called, but Leopold Martin, now 15 or so years old, showed him through to the studio. He was

… an elderly, portly, fine-looking man … very anxious, before leaving England, for an introduction to my father, and for permission to inspect his paintings. The one on the easel at that moment was the important work 'The Deluge'. The visitor, after expressing his regret and disappointment at the absence of my father, placed a chair in front of the easel … and continued for a lengthened period to gaze, without word or remark, seemingly wrapt in thought. At length he rose with the exclamation 'Mon Dieu!' at the same time taking a small bouquet from his button hole, placing it on his card, and depositing both on my father's palette. He took his departure without another word … With some curiosity I returned to the painting room, wondering who the visitor could be. Think of my delight! The card was that of the Baron Cuvier! [23]

The flooding theme turned out to have been prophetic in more ways than one. In 1827, with 300 feet of the Thames tunnel cut, the Brunels decided to take a risk and allow visitors to inspect the works for a shilling a head. Marc in particular was concerned, or perhaps paranoid, that such visits would precipitate some divine retribution for his hubris and bring about disaster. The project had proved vastly more difficult than expected; at every new inrush of water from the tunnel walls the navvies panicked and fled. But with costs mounting, more revenue had to be brought in and Brunel was under pressure from the project's directors to offset their investment. The venture was initially a success, as more than five hundred people a day came to see the works.

On May 18th they were visited by Lady Raffles (wife of Sir Stamford Raffles, the founder of Singapore) and a party of her friends. Marc was almost overcome with a presentiment of disaster, and at seven o'clock that evening the deluge came, as a spring tide swept up the Thames and breached the tunnel's roof. Isambard was in the tunnel with 160 workers when the catastrophe struck. By extraordinary fortune and great bravery on the part of the men, not a single life was lost.

The Promethean message was not lost on the curate of Rotherhithe parish church who, two days later, preached a sermon warning that the inundation was 'but a just judgement upon the presumptuous aspirations of mortal men'.[24] The curate seemed to be in the right of it, as operations to clean out the workings and stem the ingress of water from above through a pocket of silt proved both hazardous and at first ineffective. Young Isambard went down in a diving bell many times to inspect the hole in the bed of the river from above. Ton after ton of clay in bags was lowered over the hole in an attempt to staunch it. As the weeks passed, public and private confidence in the tunnel collapsed; doomsayers had a field day. After two months, with the stress and fatigue taking its toll on Marc's constitution, Isambard took more or less complete control of the recovery operations and was finally able to reach the shield which he found, fortunately, still intact.

In November 1827 it seemed that work on boring might soon resume. To restore confidence in the tunnel Marc held a grand banquet under the river for 50 friends, with the tunnel arches lit by his own patent liquefied-gaslights (the idea was borrowed from Faraday's work on gas liquefaction). The band of the Coldstream Guards provided music, and the Brunels' friend J. M. W. Turner was on hand to portray the scene in his impressionistic and evocative painting *Vaulted Hall*.[25] Toasts were drunk not only to the tunnel but also to the Royal Navy, whose victory over the Turks at Navarino (in truth not very glorious) had just been announced, posthumously sealing Byron's dream of Greek independence. The victor, Admiral Codrington, had been a staunch friend of Marc Brunel in his days in the debtors' prison.

Leopold Martin does not mention the tunnel or the banquet in his memoir, so it seems certain that he at least was not there. Whether John was present is not recorded, but he must surely have visited the works during construction. The tunnel theme is often present in his *Paradise Lost* series and, in turn, the younger Brunel was influenced by John's paintings when designing the architecture for his bridges and railways tunnels in the 1830s. A decade later he ensured that John was present for the fastest train-ride of the era. In the meantime, however, November's optimism was shattered by a second great inundation of the tunnel in January 1828, which killed six of the workmen and very nearly cost Isambard his life. Work did not resume for seven years.

One of the problems faced by workmen in the tunnel, the noxious fumes produced by rotting filth in the river water, now concentrated minds on the river itself. The Thames was the main drain for London's population. It received, noted a committee report of 1836,

> ... the excrementitious matter from nearly a million and a half of human beings; the washings of their foul linen; the filth and refuse of many hundred manufactories; the offal and decomposing vegetable substances from the markets; the foul and gory liquid from slaughter-houses, and the purulent abominations from hospitals and dissecting-rooms, too disgusting to detail.[26]

The idea that the inhabitants of Europe's greatest city were drinking this filth was an embarrassment to a generation of enterprising visionaries. At the time John Martin thought it perfectly natural that he should turn his attention to it. In 1827 he published his first pamphlet: *Plan for Supplying the Cities of London and Westminster with Pure Water from the River Colne.* This was the result of much walking and drawing and a visit paid to the River Colne with Leopold and an American engineer who possessed a new device to measure the flow of water in rivers. John's idea was to divert the River Colne by aqueduct to a great reservoir at Paddington from where, at 80 feet above the Thames, fresh water might be distributed without pumping. The surplus would be used to create a giant public bathing-place from where it would descend via waterfalls into the Serpentine River and eventually into the Thames at Whitehall via Buckingham Palace.

A second pamphlet, published the following year, was entitled *Mr John Martin's Plan for Supplying with Pure Water the Cities of London and Westminster, and of Materially Improving and Beautifying the Western Parts of the Metropolis.* The project was no mere fancy. It was accompanied by etchings of proposed waterfalls, dams, filters, water-wheels and quays. John had always wanted to be an engineer. Now he felt he could respond to both his own calling and to an urgent public need at the same time. It would be his Promethean vision come true. Or it would if anyone let him build it.

Before engineering schemes came almost exclusively to dominate his professional life for the best part of ten years, John Martin produced one more great apocalyptic canvas, *The Fall of Nineveh*, exhibited on its own at the Western Exchange in Old Bond Street in 1828. The gallery charged a shilling a head for admission. Asking the enormous price of 2,000 guineas John, even with his present fame, must have shown it less with a view to selling the painting than advertising his proposed engraving of it. This is John's own description of the subject from the exhibition catalogue: it leaves little doubt of his intention to reflect the mood of the day ...

> The moment is that in which Sardanapalus, with his concubines, is going to the pyre which, by his orders, has been erected for his and their final destruction. The oracle, which foretold the fall of the city when the river should become the enemy, had been fulfilled; and in its accomplishment he thought he beheld the manifestation of the immediate vengeance of Heaven. He had bravely resisted the human foes but against the Gods he felt that opposition must be in vain. His city flames beneath the bolts of heaven; the enemy in a flood pours through the gaping walls ...[27]

The time had long since passed when John Martin's works could be ignored by the social and aesthetic arbiters of the day, whatever the critics might say. Even before its public unveiling *Nineveh* was admired by, among others, Joseph Bonaparte, the former titular King of Spain. He presented John with a pair of Cellini candlesticks which had belonged to his brother the Emperor and which had probably been looted from Madrid somewhere around the time when Jonathan Martin was watching the retreat from La Coruña. The two-and-a-half thousand people who paid to see *Nineveh* included Sir Walter Scott, Earl Grey, a number of foreign ambassadors, Sir Thomas Lawrence and many Royal Academicians. John's failure to achieve membership of that august but conservative body was by now a matter of irrelevance.

CHAPTER ELEVEN

Playing with fire

IN SEPTEMBER 1825 *Locomotion Number One*, Robert and George Stephenson's new locomotive built specifically for the opening of the Stockton to Darlington railway, was delivered by horse-drawn wagon to Heighington Lane in Shildon, County Durham. She had been constructed at the world's first locomotive factory on Forth Street, Newcastle, managed by George's son Robert. Robert had recruited Timothy Hackworth, the talented engineer responsible for William Hedley's Wylam engines, to work at Forth Street and here they endured the birth pangs of the railway industry. The significance of the event was echoed by the publication that year of Nicholas Wood's *Practical Treatise on Rail-roads*, which became a sort of manifesto for enthusiastic railway entrepreneurs over the next decade.

At Shildon *Locomotion*, the first baby born of the union between Hackworth and the Stephensons, was carefully lowered onto the new wrought-iron I-shaped rails recently patented by John Birkinshaw. No one seemed to know quite what to do next. So they laid a fire in the firebox, filled her boiler with water and one of the navvies, John Taylor, was sent to Aycliffe for a candle and lantern so that she could be fired up. Another navvy, Robert Metcalf, thought the occasion worth celebrating with a fresh pipe of tobacco. Being an old-fashioned sort he always carried with him a pipe-glass, a little magnifying lens with which he could light his pipe using the rays of the sun. He enjoyed a few quiet puffs while the assembled men waited for Taylor to return and then, seeing a spare wad of oakum packing lying around, thought he might just kindle the engine's boiler himself with his glass. And so it was that the first locomotive to be fired on the world's most celebrated railway was lit with the celestial flame stolen by Prometheus and given to man so that he might liberate himself. It is a matter of fortune that Robert Metcalf belonged to a new breed of navvy: he could read and write, and his diary survives.

The idea that the opening of this railway heralded a new era is not just retrospect. At the time, Darlington's Quaker financier Edward Pease believed it signified an age in which the whole of the country would be connected with railways.[1] Many thousands of people were drawn to the line for its opening on September 27th, and not just to get a look at *Locomotion* under steam. For the first section of line from Witton Park Colliery to Etherley ten loaded wagons were drawn by horse, the incline being too steep for a locomotive to haul. At Etherley the horses were unhitched and the wagons drawn up an even steeper hill by a stationary steam engine connected to a winch and cable. From the top of the hill they ran unaided gently downhill to the east and at the bottom of the slope were rehitched to fresh horses. After running over the world's first iron railway bridge (exactly 100 years after Ralph Wood built the first stone railway bridge not far away at Causey Arch) the wagons were hauled by another stationary engine up Brussleton Bank while hundreds of people held on to their sides.

It was on the east side of Brussleton Bank at Shildon that *Locomotion*, crewed by George Stephenson and his brother James, lay in panting anticipation. Here, 21 new coal wagons specially fitted with seats were coupled to the train which now held 300 seated passengers and probably the same number again clinging on for dear life. After a few mishaps which included a derailed carriage and a sticky valve, *Locomotion* rolled into Darlington where she was met by a crowd of twelve thousand cheering people. At Stockton Quay, eight miles further on, she was greeted with an 18-pounder cannon salute and a crowd of more than forty thousand singing 'God Save the King'. It was a triumph for Stephenson's bloody-minded empirical skills as a project manager – a profession not of the nineteenth, but of the twentieth century.

Even as the Stockton and Darlington railway worked through its teething problems (the locomotives were not sufficiently reliable and were frequently taken out of service) George Stephenson began to pursue a much more ambitious plan to build a railway between Manchester and Liverpool. Not only was the engineering conceived on a grander and more challenging scale, involving the crossing of the notorious peaty expanse of Chat Moss and a number of difficult tunnels and cuttings; there was also violent opposition from local landowners and those with interests in the existing Bridgewater canal. Stephenson was physically threatened. Guns were fired through the night to warn off his surveyors; some of them were beaten and many of them proved in any case to be incompetent. When the bill for the new line was first presented to Parliament in spring 1825 it was torn to shreds by the enemy's counsel.

Stephenson's own appearance did not help him. He was self-consciously ill-educated and gruffly overawed in the presence of so many well-dressed, learned gentlemen. Leopold Martin later described him answering Members' questions '... with the appearance of a working-man in his Sunday best – blue coat, buff waistcoat, drab trousers, and such a white tie, wound two or three times round his neck!'[2] Disastrous as his appearance in front of the Select Committee was, Stephenson's blunt style earned him everlasting parliamentary fame when he responded to one of the members who asked him, would it not be very serious and awkward if any obstructions – a cow, for instance – got upon the railway in front of the engine. 'Yes,' he replied, 'varry aakward – for the coo.'[3]

Stephenson was an object of much interest among London's chattering classes. To some he was an object of patronizing curiosity; to others he embodied the fascination of an exotic beast. From the diary of the young and beautiful actress Fanny Kemble (niece of the Mrs Siddons who had visited John Martin's studio and posed as the prophet Daniel) there is a droll study of the engineer in an exuberant and perceptive letter dating from her first ride in one of his locomotives:

A common sheet of paper is enough for love, but a foolscap extra can alone contain a railroad and my ecstasies. There was once a man, who was born at Newcastle-upon-Tyne, who was a common coal-digger; this man had an immense constructiveness, which displayed itself in pulling his watch to pieces and putting it together again; in making a pair of shoes when he happened to be some days without occupation; finally – here there is a great gap in my story – it brought him in the capacity of an engineer before a committee in the House of Commons, with his head full of plans for constructing a railroad from Liverpool to Manchester. It so happened that to the quickest and most powerful perceptions and conceptions, to the most indefatigable industry and perseverance, and the most accurate knowledge of the phenomena of nature as they affect his peculiar labours, this man joined an utter want of the 'gift of the gab'; he could no more explain to others what he meant to do and how he meant to do it, than he could fly …[4]

Stephenson returned to the north nursing the wound to his pride but with no intention of giving up on his scheme. He was intensely competitive, and the thought of another engineer succeeding where he had so far failed was too much to bear. The sharp and talented younger Stephenson, Robert, was not on hand to help his floundering father through this dark period. In wishing to assert his growing independence he had accepted a three-year post as a mining engineer in Colombia. In his absence the engine manufactory at Newcastle was directionless and relations with the Stockton and Darlington shareholders became strained. But despite these problems and much disparaging criticism from other engineers, George stuck to his guns and the new line slowly, very slowly, took shape.

The South American project proved impossible with the technology available to Robert. He returned to England in 1827, at the same time ensuring with a gift of £50 that the stranded Richard Trevithick, whom he met in a bar in Cartagena, might take passage home too. Immediately the family fortunes were revitalized by Robert's energy, business acumen and engineering expertise. Like the younger Brunel who, during the long interruption to the Thames tunnel project had also become interested in railways, Robert had the benefit of being apprenticed to a father who had learned his trade the hard way: through empirical trial and error and in the absence of recognized engineering or professional standards. The Liverpool to Manchester contracts, for example, barely existed on paper: George made many of them up as he went along. The second-generation engineers, the sons, had the benefit not only of their fathers' talent and experience; they also grew up in an environment in which standards of professionalism were becoming widely accepted, and in which there were sufficient rivalries to ensure that only best practice succeeded.

Perhaps the most surprising battle that the early locomotive engineers faced was the need to demonstrate that the future of transportation lay with the locomotive and not the competing array of technologies which, now obsolete, seem to modern eyes to have been doomed from the start. George Stephenson had already shown on the Stockton and Darlington line that a combination of horse, stationary engine and locomotive was viable. His early designs for the Liverpool to Manchester line provided for a similar combination.

By the winter of 1828 the debate was at its most bitter and vested interests seemed to have swung in favour of mixed traction. But Robert Stephenson, now working hard on improving the performance of his locomotives, was determined that they should win the day. 'Rely upon it,' he wrote, 'locomotives shall not be cowardly given up. I will fight for them until the last. They are worthy of the conflict.'⁵ He prevailed. In April 1829 a resolution by the directors of the proposed new line was passed offering a prize of £500 and, in effect, the contract for the Liverpool and Manchester route, to the best locomotive, the trials to be held at Rainhill, near Liverpool, in October. Now Robert Stephenson set to work building a locomotive to win the prize and prove the technology, implementing ideas that he had been playing with since his days in South America. The result, named *Rocket*, beat all-comers for speed and reliability and won for the Stephensons the right to complete their railway. Against competition from mixed traction, from steam carriages and from horses, *Rocket* proved that locomotive technology was the way forward.

Jonathan Martin made no mention in his autobiography of the dramatic and historic events which had been played out on his doorstep in Darlington during the mid-1820s. Probably he believed that this new technology with its belching steam, clanking wagons and godless, swearing drivers and firemen was the work of Satan. And the locomotive was not the only ungodly invention in the neighbourhood. In 1827 a Stockton inventor, John Walker, patented a friction match which he called the Lucifer. Michael Faraday exhibited it shortly afterwards during a lecture at the Royal Institution.

Jonathan's Promethean paranoia was at this time exhibiting itself in horrifying dreams in which the son of Napoleon Bonaparte invaded England and held her capital to ransom. Babylon-on-Thames, so Jonathan believed, would be burned unless its sinning inhabitants turned at once to the word of God. William made a primitive engraving of the dream for the second edition of Jonathan's *Life*; evidently, then, the two were in touch. Their mutual sympathy may have grown stronger during these years as William's own mania developed. In his case it was a conviction that Isaac Newton was a false prophet. He had recalculated Newton's gravitational constant and found it to be out by a factor of 800. Newton had, in his view, been fundamentally wrong in his clockwork, mechanical interpretation of the universe. William had already demonstrated to his own satisfaction with his perpetual motion machine (still running unaided) that the Secondary Cause of all things was air. Now he developed this thesis (much to John's embarrassment) into a Martinian System of philosophy which he proclaimed in lectures and broadsheets from 1827 more or less until his death 24 years later.

Jonathan had by now given up, or lost, his employment in the tannery at Darlington. He spent much of his time preaching as a Primitive Methodist and hawking his autobiography. His son Richard, whom he had managed to look after for some years, he now 'gave' to a Jewish street trader as an assistant. In 1827 Jonathan went to Lincoln and after getting into more trouble preaching fire and damnation to the people there, found employment again as a tanner. Frequent absences from work – either for preaching or selling a new, third edition of his *Life* – lost him this job too and he took to the road again, selling his booklet to those he could interest or who took pity on him. He was not shy of exploiting the name of his famous brother John, who had just been given a diamond ring and gold medal by Nicholas I of Russia, to whom he had dedicated *The Deluge*.

During these ramblings Jonathan found time to pay court to a woman from Boston in Lincolnshire, called Maria Hudson. They married in 1828 and returned to Lincoln, where Richard came to lodge with them. On Boxing Day that year, leaving Richard behind in a Lincoln boarding school, Jonathan and Maria left for York. The following day Jonathan wrote this grim and prophetic warning, the first of several, to the clergy of that city:

> I right Oh Clargmen to you to warn you to fly from the roth to
> cum you who are bringing a Grevus Cors upon the Land you blind
> Gieds and Decevers of the Peopeal How can you Easpe the
> Damnation of Hell you whitent Sea pulkirs you who are Draging
> Millians of Souls to Hell with you will not the Rich and the myty
> have to Curs the Day thay sat under your Blind and Halish Doctren
> but I warn you to repent and cry for marcy for the Sorde of Justes
> is at Hand and your Gret Charchis and Minstaris will come rattling
> down upon your Giltey Heads for the Sun of Boney part is prepar-
> ing for you and he will finish the work his Father has left undun.
>
> Jontan Martin
> Your sinsear Frind [6]

Jonathan went into York Minster a day or so later and impaled the letter on an iron spike of the gate which separated the south aisle of the great church from the choir. A brushmaker called William Scott noticed it, took it down and kept it, but without informing anyone of its contents.

Jonathan Martin's may have been one of the more extreme expressions of admonishment, but it was not the only one that year. Thomas Carlyle, the reactionary Scottish Calvinist, was moved to write a piece for publication in the *Edinburgh Review* entitled 'Signs of the Times'.

Like Jonathan and many more rational conservatives he was intensely discomfited by what he saw as the end of a golden age of human morals and creativity, and the arrival of a less intuitive, more mechanical age. He made this stern observation:

> These things, which we state lightly enough here, are yet of deep import, and indicate a mighty change in our whole manner of existence. For the same habit regulates not our modes of action alone, but our modes of thought and feeling. Men are grown mechanical in head and in heart, as well as in hand. They have lost faith in individual endeavour, and in natural force, of any kind.[7]

Carlyle could not help noticing that along with the traditional artisan, modern society was also casting aside an intimate, homespun relationship with nature and with God:

> For all earthly, and for some unearthly purposes, we have machines and mechanic furtherances; for mincing our cabbages; for casting us into magnetic sleep. We remove mountains, and make seas our smooth highways; nothing can resist us. We war with rude Nature; and, by our resistless engines, come off always victorious, and loaded with spoils.

The Promethean implications of this hubris, articulated in a way that the frustrated and maniacal Jonathan Martin could not himself express, were equally obvious to Carlyle: 'Our true Deity is Mechanism. It has subdued external Nature for us, and we think it will do all other things. We are Giants in physical power: in a deeper than metaphorical sense, we are Titans, that strive, by heaping mountain on mountain, to conquer Heaven also.' Comparing the present to the follies and atrocities of the French Revolution Carlyle admitted frankly that Old England's maladies had grown milder in recent times. He nevertheless saw a bleak future and had this warning for his middle-class readers: 'The King has virtually abdicated; the Church is a widow, without jointure; public principle is gone; private honesty is going; society, in short, is fast falling in pieces; and a time of unmixed evil is come on us.'[8]

His opponents did not allow Carlyle and the conservative establishment to peddle their rhetoric unanswered. The Prometheans, united by the zeal of insolent liberation and reformism, were now mature and self-confident. Turner celebrated the new age in paint because in his hands colour was a medium with infinite potential. John Martin was attracted to mechanical and artistic innovation with a childlike sense of novelty. Davy and Faraday had seen in chemistry a key to nature's secrets. Had Shelley been alive he would surely have unsheathed his sword and challenged Carlyle. Without him it was left to another angry young man, Charles Babbage, to raise the Promethean standard in print.

Babbage was the son of a Devonshire banker and in this respect, as in many others, was not typical of his fellow Prometheans. He was an unusual sort of mathematician, too. He was socially gifted and in much demand among scientists, industrialists and wealthy patrons. His house at Number 1, Dorset Street (formerly belonging to Dr Wollaston) became a centre for lively gatherings of the Marylebone set which included Faraday, Wheatstone, the Martins and many other sympathetically creative types. He married for love and happily so until the death of his wife Georgiana in 1827. Even more remarkable was his determination to apply science both to real engineering issues and to politics and economy. He spent many months touring the manufactories of Britain and knew their workings intimately. His liberal scientific views led to his involvement in the high-stakes politics of the Reform Bill.

While at Cambridge, Babbage had become interested in computers: the people who calculated numerical tables for actuaries, mathematicians and government departments. The tables produced by tedious and lengthy manual arithmetic were notoriously prone to error. Babbage once remarked casually to his friend John Herschel (son of the astronomer Sir William), when they were struggling with some of these tables, that he 'wished to God these calculations had been executed by steam'. Herschel replied that the idea was 'quite possible'.[9] In 1822 Babbage wrote to Sir Humphry Davy that he believed he could construct a mechanical calculator that would replace the 'computers' and shortly thereafter began to design the Difference Engine. It would require 25,000 parts and weigh 13 tons. It would perform calculations using the method of finite differences and compute to an accuracy of 31 digits. Davy's response was encouraging and, with the support of the Duke of Wellington, Babbage began to conceive how he might carry his plan into action.

His problem was to find an engineer capable not only of realizing his designs, but also of designing the machinery to construct the tools with which to make the parts for the engine. It was a monumental task, made more difficult by the fact that having conceived his Difference Engine #1, and an improved design, Difference Engine #2, he went on to conceive a much grander, much more complicated machine which to all intents and purposes would have functioned like a modern digital programmable computer. He called it the Analytical Engine.

Against this background, in 1830 Babbage published a paper entitled 'Reflections on the Decline of Science in England and on some of its Causes'. Much of it was an attack on the Royal Society and in particular its President, Davies Gilbert, but it also stands as a manifesto for scientific method, for progress, application and perfectibility. As such it owes much to William Godwin and his fellow Sandemanian Michael Faraday. If Carlyle and the reactionary generation which he represented feared change, Babbage feared stasis, the gradual corruption through apathy of the Promethean spirit which had brought about the genesis of a modern world.

Babbage, a more or less exact contemporary of Shelley and of Michael Faraday, was in one sense a throwback to the previous generation of empirical amateur talent. He never made any money. He had the mind of a Coleridge or an older Brunel (of whom he was a good friend): extraordinarily fertile, too clever almost for anyone to understand him. He had been too clever as a student. Failing to gain a degree, by 1828 at the age of 39 he had nevertheless been made Lucasian Professor of Mathematics, the post that Isaac Newton had held at Cambridge. He learned of the appointment while in Rome, reading the announcement in a newspaper. It has taken 150 years to replicate the extraordinary calculating machines that he now attempted to build.

By criticizing the Royal Society Babbage incurred the opposition of his friend Michael Faraday, who felt constrained to reply to him in print refuting his accusations. In the end, Gilbert resigned the Presidency, such was Babbage's standing in the European community of scientists. His landmark work of 1832 *On the Economy of Machinery and Manufactures* is cited as a formative influence on both John Stuart Mill and Karl Marx.[10]

Of more than passing interest in this context is a passage in the 'Decline of Science' which exemplifies Babbage's view of scientific method. It also underlines the brilliance of Faraday's careful, diligent approach to experiment which made him the greatest scientist of the nineteenth century. It is an anecdote almost too good to be true:

> Meeting Dr Wollaston one morning in the shop of a bookseller, I proposed this question: if two volumes of hydrogen and one of oxygen are mixed together in a vessel, and if by mechanical pressure they can be so condensed as to become of the same specific gravity as water, will the gases under these circumstances unite and form water?
>
> 'What do you think they will do?' said Dr W. I replied, that I should rather expect they would unite. 'I see no reason to suppose it,' said he. I then inquired whether he thought the experiment worth making. He answered, that he did not, for that he should think it would certainly *not* succeed.
>
> A few days after, I proposed the same question to Sir Humphry Davy. He at once said, 'They will become water, of course'; and on my inquiring whether he thought the experiment worth making, he observed that it was a good experiment, but one which was hardly necessary to make, as it must succeed.[11]

Michael Faraday would not only have disagreed with Sir Humphry that the experiment was not worth making; he would have conducted it himself the same day. At this time he was, in fact, preoccupied with sound and the optical properties of glass and within two years of making the breakthrough which established his undisputed pre-eminence as an investigator. He was also much in demand as a lecturer and had been appointed director of the laboratory at the Royal Institution on Davy's recommendation. The former mentor had finally, and generously, become reconciled to the talents of his pupil. It was just in time: Davy died in Geneva in 1829 at the age of 51: a victim, in the end, of his earlier reckless disregard for safety in the laboratory.

At the beginning of 1829 John Martin too was preoccupied. Not content with redesigning London and mastering steel mezzotint, he had tried to interest the authorities in various useful inventions, such as a 'laminated beam', an 'elastic iron ship', 'coast lights on a new principle' and a plan for ventilating coal mines.[12] Painting, it seemed, had begun to bore him; his imagination, like that of his brothers, must now be played out in real life. He had returned to London after a visit to William in Newcastle and it is not surprising that he was enthused by such schemes: it was a family trait. But before the New Year was six weeks old he had much more pressing matters on his mind.

Jonathan's letter to the clergy of York Minster had been followed by others. One he tied by a string to the same railings as the first, next to the choir. It was read by a verger and passed by him to the assistant vicar-choral, but no action was taken. Jonathan was not the only religious zealot in early nineteenth-century England; how was one to know which threats were genuine? Another note, in a similar vein to the others, was picked up by a choir boy. A fourth was addressed to 'You blind Hypacrits, you Saarpents and vipears of Hell, you wine Bibears and beffe Yeaters'.[13] On the 21st of January, furious that his requests (or what he evidently conceived were requests) for a debate with the clergy had fallen on deaf ears, he wrapped this letter and a copy of his *Life* around a stone and tried to throw it through one of the stained-glass windows of the Minster to attract attention. It bounced back, so he carried it into the south aisle himself. He did not, apparently, see this as a divine warning not to attack the Minster, but as a provocation. The parcel having been picked up by a sailor from Hull and taken as a souvenir, Jonathan again failed to receive the reply that he sought. In desperation he prayed for guidance.

> The next night I dreamt that a wonderful thick cloud came from
> heaven, and rested upon the Minster, and it rolled towards me at
> my lodgings – it awoke me out of my sleep, and I asked the Lord

what it meant – and he told me it was to warn these Clergymen of England, who were going to plays, and cards, and such like – and the Lord told me he had chosen me to warn them – and reminded me of the prophecies – that there should in the latter days be signs in the heavens … I felt a voice inwardly speak, that the Lord had chosen me to destroy the Cathedral.[14]

Jonathan now saw his destiny calling: to act out, in real life, the sublime apocalypse of his brother's fantasy canvases. His greatest worry was that his new wife might try to stop him from doing the Lord's bidding. So he took her wedding ring from her while she slept and when she woke promised to return it to her only if she would keep secret what he was about to tell her. Maria was appalled to learn that Jonathan had taken upon himself the task of fulfilling the Lord's vengeance against the clergy. She managed to persuade him to leave York and go with her to Leeds where distance might calm his morbid temptation. For days he struggled with his conscience while Maria implored him not to abandon her or his son Richard, still at school in Lincoln. But the voices in Jonathan's head became more persistent with every day that passed and eventually, on the last day of January 1829, he took his leave of her and returned to York. He arrived there at 10:30 the next morning.

It was a Sunday. He hung around the Minster all day, increasingly irritated by what he called the 'buzz' of the organ. At the end of the evening service he secreted himself behind the monument to Archbishop Greenfield in the north transept and by half-past six, when the sexton locked the doors of the Minster believing it to be empty of people, Jonathan found himself alone in pitch darkness and freezing cold.

He fumbled his way to the stairs which led up to the bell chamber in the tower at the southwest corner of the Minster, and there afforded himself the luxury of a light using a razor and flint and half a penny candle. By chance 1829 was the year in which a London chemist, Samuel Jones, patented a match which he called The Promethean, but Jonathan evidently preferred older, trusted methods.

Now he cut himself a hundred feet of bell rope which he knotted in a seamanlike manner to make a rope ladder. He put out his candle to avoid being seen from outside while he was at his business, and descended to the nave with the rope. He then used it to climb over the railings into the choir where his intended target was the throne of the archbishop, the very seat of Anglican diocesan power in the north of England, dating back to the days of King Edwin in the early seventh century. The throne was lavishly curtained in silk-lined crimson velvet, which Jonathan cut down to make himself a sort of robe which he decorated with gold tassels from the pulpit cushions. Now he collected together prayer books and cushions from the benches in the choir and formed two heaps of them – one between the throne and the choir screen and another against the stalls on the north side of the choir. He knelt and prayed, thanked the Lord for guidance and cried 'Glory be to God!'

Jonathan Martin lit his two heaps of kindling and, throwing his improvised ladder over the railings, climbed back over them and down into the nave. He managed to locate a mobile scaffolding tower used by workmen to reach the Minster's windows and scaled it like a true foretopman, breaking a window with a pair of pincers he had stolen beforehand and letting himself down to the ground outside with the rope ladder. There had been a sharp frost and there was ice on the pavement. He fell to his knees and gave thanks, and the Minster clock struck three in the morning.

Thomas Sopwith, a pioneering Newcastle geologist who at one time or another knew all the Martin brothers, wrote when he heard the news:

> Poetic imagination can hardly conceive a more distressing or remarkable scene than this poor idiot wandering alone in the vast aisles of the glorious structure, the last and only spectator of that magnificent choir, on which the beams of light had shed their parting rays, and the chords of the organ had sounded their last rolling thunders and sweet melodies never to be again heard.[15]

Bizarrely, no one, not even a couple of early passers-by who had seen odd lights in the Minster, reported the fire until seven o'clock that morning when a choirboy, arriving early for his lessons, slipped on the ice and on getting up noticed smoke coming from the window through which Jonathan had escaped. He rushed to report it and was soon joined by a party of Minster masons and workmen whose yard lay just across the road.

When the sexton and his party got to the scene of the blaze they found the throne and the south choir stalls well and truly alight and the fire spreading. Some of them were put to rescuing furnishings and the great eagle lectern, while the Minster's fire engine was brought from the vestry. Smoke began to hamper their efforts and the call went out for the city's fire engine which, when it arrived, played more or less uselessly on the spreading flames. By eight o'clock that morning the fire had reached the organ screen, from where it very quickly enveloped the organ, thought to have been the finest in England, together with precious musical scores. The organ responded by playing ghostly tones as it burned, quite unnerving the crowds now gathering outside to watch. Molten lead from the roof of the choir began to rain down. From afar the scene looked like a giant furnace, the flames reflected in and glowing through the Minster's unique medieval stained-glass windows. One of the watching crowd was heard to say, at the moment when the flames were at their greatest height and threatening to destroy the entire building, that the scene would have made a wonderful subject for the brush of John Martin.[16]

Only now, when the fire was obviously out of control, were messages dispatched to Leeds and Tadcaster for larger and more powerful engines to be brought. Those summoned from Leeds, 24 miles away, arrived just two hours after their crews were raised – a staggering feat of speed and endurance. Once it was realized that the choir roof must fall and that there was a very real threat to the priceless great east window, the operation was undertaken with extreme swiftness and immense bravery. But by nine in the morning, just two hours after the first alarm was raised, the fourteenth-century oak choir roof came crashing down, its fallen burning timbers now threatening the limestone pillars of the choir. Men rushed to haul the charred beams out of the Minster and into the yard outside while panicking birds and bats flitted about inside, screeching in frantic attempts to escape the smoke and flames. Eventually the 12 most powerful engines that could be deployed were brought to bear and by the end of the day the fire was under control and the east window saved.

Five days after the fire the Dean and Chapter of the Minster, having conducted an inquiry into the causes of the blaze and interviewed several witnesses, caused an advertisement to be printed in all the newspapers serving the northeast of England:

WHEREAS
JONATHAN MARTIN
Stands Charged with having on the Night of the 1st of February, Instant,

WILFULLY SET FIRE TO

YORK
MINSTER.

A REWARD OF
100 POUNDS

Will be Paid on his being Apprehended and Lodged in any of his Majesty's Gaols.

And a Further Reward of

One Hundred Pounds

Will be paid on the Conviction of any ACCOMPLICES of the said JONATHAN MARTIN, to such Person or Persons as shall give Information which may lead to such Conviction.

The following is a Description of the said Jonathan Martin: viz.

He is rather a Stout Man, about Five Feet Six Inches high, with light Hair cut close, coming to a point in the centre of the Forehead, and high above the Temples, and has large bushy Red Whiskers; he is between Forty and Fifty Years of Age, and of singular Manners. He usually wears a single-breasted blue Coat, with a stand-up Collar, and Buttons covered with the same cloth; a black cloth Waistcoat; and blue cloth Trowsers; Half-Boots laced-up in front; and a glazed, broad-brimmed, low-crowned Hat. Sometimes he wears a double-breasted blue Coat with yellow Buttons.—When Travelling, he wears a large black leather Cape coming down to his Elbows, with two Pockets within the Cape; there is a square piece of dark coloured Fur, extending from one shoulder point to the other.—At other times he wears a drab coloured great Coat, with a large Cape and shortish Skirts—When seen at York last Sunday, he had on the double-breasted blue Coat, a common Hat, and his great Coat.

The said JONATHAN MARTIN is a Hawker of a Pamphlet entitled "The Life of Jonathan Martin, of Darlington, Tanner," the Third Edition of which is printed at Lincoln, by R. E. LEARY, 1828.—He had lodged in York about a Month, and quitted it on the 27th of January last, stating that he was going to Tadcaster for a few Days, and thence to Leeds. He returned to York on the 31st of January, and said that he and his Wife had taken Lodgings in Leeds. He was not seen in York after the 1st of February.

By Order of the DEAN and CHAPTER of YORK,

CHRIST. JNO. NEWSTEAD,
Clerk of the Peace for the Liberty of St. Peter of York.

York, 5th February, 1829.

The fugitive was described as a rather stout man, five feet six inches tall with light hair close-cut and red bushy whiskers, of between forty and fifty years old and of singular manners. His normal dress was a single-breasted blue coat with a high collar and cloth-covered buttons, a black cloth waistcoat, blue trousers and a distinctive glazed broad-brimmed hat. He was known to be a hawker of an autobiographical pamphlet.

Jonathan did not want to be caught but he had advertised his intentions so obviously that he was easily identified as the incendiary and across three northern counties he was pursued by determined thief-takers eager for reward and fame. Despite being weighed down by a very substantial bundle containing copies of his *Life*, 9 feet of gold fringe, 22 feet of crimson velvet from the Archbishop's throne, a small Bible and some pieces of stained glass which he had also kept as souvenirs, he made extraordinarily rapid progress as he headed northwest. Through Easingwold and Thirsk, where he sold one of his pamphlets and drank 5 glasses of beer, he walked briskly, covering 32 miles in 12 hours. He made no attempt to disguise himself, though he hid when coaches passed him on the road, knowing that a hue and cry must have been raised.

At Northallerton he took brief refuge with his first wife's brother Joseph Wilson. Wilson, realizing Jonathan was in trouble but having heard nothing as yet of the fire, arranged for him to be taken by horse and cart to West Auckland. From here he struck out on foot to Allensford, a remote lead-mining town in the north Pennines. He spent the night there and then set out for Corbridge and the temporary safety of Edward Kell's house. Kell was a very old friend of the Martin family who had sheltered Jonathan after his escape from the Gateshead asylum. In three days the fugitive had travelled more than a hundred miles, mostly on foot and with a heavy load.

On Friday 6 February 1829 Newcastle newspapers arrived at Hexham carrying the advertisement of the Minster fire. William Stainthorpe, an innkeeper and Sheriff's officer at Hexham, knowing that Jonathan had friends and relations in the area, immediately set out to look for him, inquiring of anyone he met if they had seen the fugitive. Encouraged by hints of Jonathan's arrival in the Tyne Valley he was able to track him to Kell's house at Codlaw Hill by the end of the afternoon and there apprehended him without a struggle. Jonathan and his bundle were now escorted to Hexham, at one point passing the house in which he had been born, which he pointed out to his captor. He was passive, unresisting and perfectly open in his answers to Stainthorpe's questions. He admitted having set light to the Minster and was curious to know how much of the building had been damaged. About a hundred thousand pounds' worth, Stainthorpe thought.[17] By Friday evening word reached York by express messenger that the Minster arsonist was taken. On Monday 9 February 1829 at three in the morning Jonathan was brought secretly by private coach from Hexham Prison to York to avoid the angry crowds that were expecting his arrival.

It was not until the first or second week of March that Jonathan, who had been interviewed several times and remained unrepentant, was visited by his brothers. William came down from Wallsend, staying only a few days. Richard, now retired from the army after 29 years' service and living on Dean Street in Soho, who must have been briefed in London, stayed in York to arrange the defence case with a solicitor he had employed called R. H. Anderson. He may have persuaded his older brother to leave York because William's own monomania, though not as severe as Jonathan's, led him to a sympathetic view of the burning of the Minster, which he later celebrated in 31 verses of awful doggerel.[18]

Jonathan was appalled to find that his defence was to be that of insanity. His brothers, he felt, had betrayed him in his hour of need. It was in truth the only possible defence: arson was a capital crime and Jonathan must be hanged if he was not proved mad. John did not come north, but that is not to say that he was careless of Jonathan's fate. He had undertaken, at what must have been enormous expense, for Jonathan to be defended by none other than Henry Brougham.

In the decade since the trial of Princess Caroline, Brougham had worked hard to restore his credibility among Whigs by promoting a national system of education. He had helped to launch the much-derided but worthy Society for the Diffusion of Useful Knowledge.

He had been a prime mover in the establishment of London University and the Mechanics' Institutes. He had tried to forge reformist cross-party coalitions in the House of Commons and for the most part he had failed. But his fortunes were changing once more. The Parliamentary elections of 1826 were notoriously corrupt and, after a lacuna of several years, popular demand for reform was once again hot. With the departure of Lord Liverpool after 15 years as Prime Minister in 1827, the Duke of Wellington took the reins of government.[19]

Unable to prevent a party split over the long-running sore of Catholic emancipation, and labouring in apparent denial of widespread economic hardship, the duke presided over a government whose majority was squeezed between the tolerant left and ultra-conservative right.

Brougham once more raised his standard among the Whigs, newly united and baying for blood. He was not to become their leader, however. The party had not forgiven his many sins and George IV was naturally enough his implacable enemy; but across the country, among the 'People', he was the Whigs' populist champion. According to Macaulay he was, 'next to the King, the most popular man in the country'.[20] Within a year of Jonathan's trial George IV was dead and in the new parliament Brougham was sensationally elected for the County of Yorkshire on a platform of religious tolerance, abolition of slavery and the repeal of the Corn Laws. Earl Grey was more or less forced to appoint him Lord Chancellor in a cabinet poised to fulfil the long-cherished dreams of the reform movement.

Brougham's acceptance of Jonathan's case reflects not only a desire to keep a high public profile, but also his recognition of the religious sensitivities and political potential of the case. Jonathan's professed adherence to Methodism and the almost universal abhorrence of that sect among Anglicans exposed a raw nerve during the debate over Catholic emancipation. Politically Brougham was apt to overplay his hand; but as a legal advocate his judgement was impeccable. The case must not be allowed to become a religious cause célèbre.

The trial of Jonathan Martin in March 1829 was almost as great a national sensation as that of the notorious bodysnatchers Burke and

Hare in Edinburgh two months previously. William Burke had been hanged for the murders of sixteen people just three days before the Minster fire, in front of a crowd of twenty thousand people. His accomplice, by turning King's Evidence, escaped the noose and was set free.

Jonathan Martin's trial, having been transferred from the City Assizes to the County Assizes at York Castle, began on March 30th. Although having tried to abscond from prison, and notwithstanding his anger against his brothers for wishing to see him declared insane,

> Martin … maintained his composure, his cheerfulness and his health. Since the frustration of his attempt to escape, he has settled in the place of his confinement with a philosophical indifference. Civil in his deportment, and obliging throughout the whole of his demeanor, he appeared to those who visited him, completely free from all consciousness of having done anything wrong, or from any apprehension as to the consequences of his conduct. The smile of gratulation seemed rather to illumine his countenance, and a satisfactory reminiscence to preserve the tranquility of his thoughts. He has taken an unusual degree of exercise – daily pacing up and down the yard of his apartment, at the rate of about five miles an hour, sometimes walking as much as 20 miles per day.[21]

The irony of Jonathan's case, in which there were no grounds for doubt as to his having committed the crime (his plea was that his God was guilty and the judge entered this as 'not guilty') was that those of his friends who appeared for the defence wished it to be known that they thought him of sound mind, while his enemies had no doubt that he was insane. The chief counsel for the prosecution, Mr Alderson, was at pains to ensure that the jury understood the law concerning criminal responsibility, which in 1829 predated the McNaughton rules[22] by more than a decade:

> That you may understand it the better, I will state what I consider the law laid down upon the question of insanity, so as to affect the prisoner with the consequences of guilt. I apprehend, that in order

to make him dispunishable by law, for a criminal offence, you must
be satisfied, that at the time the offence was committed he was
incapable of distinguishing between right and wrong. However he
may have, on former occasions, been afflicted even with insanity,
still, if at the time the offence was committed, he was capable of
making the distinction I have mentioned, of knowing the consequences – of knowing he was doing a wrong act, he is accountable
for his conduct.[23]

Mr Anderson went on to cite a number of cases which had turned on
this distinction, including that of John Bellingham, the Liverpool
banker who murdered Spencer Perceval in the lobby of the House of
Commons in 1812 and been hanged for his crime.

The witnesses called by Brougham, including the decent asylum
keeper Richard Nicholson, the family with whom he had lodged in
York, and Jonathan's old friend Edward Kell, were of little help to the
defence. They all, whilst excepting his religious views, had found him
to be a decent man and of sound mind. On the other hand Mrs Orton,
Jonathan's second and much less sympathetic keeper at Gateshead,
had no doubt that he was insane. These testimonies can only have
served to confuse the jury. Jonathan's own account, on the other hand,
was unwittingly of great help. He spoke very rapidly in his broad
Northumbrian dialect and related the whole story of his dreams, the
voices telling him that he was God's avenger against the clergy, and
how he had carried his scheme into practice. Frequently throughout the
trial he interrupted witnesses and counsel alike to correct them on
matters of fact. A modern clinical psychiatrist would very probably
have diagnosed Jonathan with some form of schizophrenia.

The last witnesses called by Henry Brougham were decisive. These
were three doctors who had examined Jonathan. The first, a Newcastle surgeon who had attended him at the Gateshead Asylum, declared
that the defendant was a monomaniac. Two other doctors, who had
seen Jonathan several times in the gaol at York, agreed. One of them
declared that he had detected an injury to Jonathan's skull, presumably one caused during his many adventures at sea. Brougham may

have reminded the jury that James Hadfield, George III's would-be assassin, had suffered from a head injury which was thought to have caused his insanity. It was a persuasive point. When the prosecution suggested that Jonathan's flight and attempts to evade capture showed evidence of conscious guilt, all three doctors disagreed.

The trial, having been carried over to the 31st March while extra witnesses were prepared, lasted a single day. At the end of the afternoon the foreman of the jury told the court, 'We are of the opinion that he set fire to the Cathedral, being at the time insane or of unsound mind.' Jonathan Martin was sentenced to be detained at His Majesty's pleasure.

On 29 April 1829, at the age of 47 (he gave his age as 42) Jonathan was received at London's Criminal Lunatic Asylum, the Hospital of St Mary of Bethlehem: Bedlam.[24] He arrived with something of a reputation, having tried to escape from the gaol at York by climbing up a chimney there. The gaoler, John Kelly, reported to the authorities at Bedlam that Richard Martin had told him 'his brother is very fond of liberty'.

Curious visitors found Jonathan perfectly harmless unless they unwisely broached the subject of religion with him. He occupied a very plain whitewashed room on the first floor of the building, with a barred window high up in one wall. There was a table, bed and chair, all kept very clean. His presence in the hospital put at least one nose out of joint. Since the death of Margaret Nicholson, the woman who had vainly tried to stab George III with a dessert-knife in 1786, the most notorious inmate at Bedlam had been James Hadfield, who had shot at the king at the Drury Lane Theatre in 1800. Hadfield was an army veteran with a good service record whose behaviour had changed after he received a wound to the head. Liberal reformers arranged for him to be defended by the most brilliant advocate of his generation, Thomas Erskine, who argued successfully that Hadfield's 'irresistible impulse' was a lunatic delusion, an act for which he was not responsible.[25] It was as a result of his trial that the Criminal Lunatics Act had been passed the same year. This was the act which had saved Jonathan's life. Now that a new celebrity was drawing all the attention, Hadfield (who had made one successful escape from Bedlam and been captured at Dover) was most unhappy.

Occasional fits of monomania resulted in Jonathan being restrained by manacles. At such times, but more usually when his behaviour was good, Jonathan was given paper and inks to draw and paint with. These may have been supplied by his brothers John and Richard, who came to visit him periodically, by his son Richard, who went to live with John and Susan in Marylebone, or perhaps by his sister Ann, who also had a room at Allsop Terrace. Jonathan may, in fact, have been the prototype art therapy patient, for his doctors were aware that drawing soothed his mania and allowed him a form of self-expression. His drawings form almost the earliest part of the hospital's internationally renowned collections which include works by, for example, the parricide Richard Dadd. That they are the work of a disturbed mind there is little doubt.

The reverse of a picture called *London's Overthrow*, made in 1830, is covered with an immense, minutely drawn series of calculations which Jonathan made to show how much money the government owed him for false imprisonment. Jonathan's compulsion to draw and paint is shown by his own inscription describing this picture's creation:

> When I began this picture for to draw the devil kicked it with his foot, and my mess bowl at it did throw … he forced me from the table, and turned me to the door, then on the seat I placed my plan surrounded by the worst of men, then the wind began to blow, which did my paper overthrow, then with my slates and books I held it down, with a pound of iron on each hand, bound to my loins with seven pounds more, upon one knee, dauntless my painful task pursued like those valiant Jews of old …[26]

There is no doubt that *London's Overthrow* was inspired by John's *Fall of Nineveh*, of which Jonathan must have been given a mezzotint copy by his brother. It is an extraordinary vision. To the right, stretching from the foreground into the distance with the full force of a John Martin perspective, lies London on fire, a lion's head visible through the smoke. Saint Paul's Cathedral is enveloped in smoke and one wonders if the novelist William Harrison Ainsworth was inspired by word of this image when he recreated Jonathan in his apocalyptic prophet Solomon Eagle. In the foreground, priests, merchants and other assorted sinners repent too late. High on a hill to the south of the Thames, overlooking the scene of his triumph across the river, a very decent caricature of Napoleon Bonaparte sits astride his horse, surrounded by booming cannons. Jonathan's handwritten narration at the bottom of the drawing quite unnecessarily reinforces his dramatic message:

> Behold blackness has covered every face and a voice is heard from the midst of the people and lamentation and weeping; now is babylon fallen, she has fallen, she that hath made the merchants rich with her merchandize and the seamen standing afar off clapping their hands together and crying Alas! That great city that made us all rich in one day is burnt to the ground.

One visitor who came to see Jonathan three years after his arrival was a journalist, unfortunately anonymous, who wrote a piece about him for Tilt's *Monthly Magazine* which was published alongside an article on William Blake. The writer found Jonathan in the middle of making a picture which depicted a bishop with seven heads being swallowed by a crocodile. The lunatic seemed to him perfectly harmless in appearance, with a smooth and unruffled face showing no sign of guilt. He was evidently disappointed with this real-life firebringer:

> His manners were modest and rustic: he conversed like a good-humoured cottager, not overburdened with sense, nor made sulky by circumstances. He was not merely resigned, but perfectly satisfied: he wanted nothing but lots of Indian ink and Brookman's black-lead pencils. Although in his way a reformer of the church, he knew nothing of Wickliffe; he had never heard of Luther or Calvin. He was, upon principle, the antagonist of episcopacy, and he thought the best way of pulling down the priests of Baal was to burn them out of their high places. I had made him out in my mind's eye a Prometheus. I expected to have found him sublime; but he was quite insignificant.[27]

CHAPTER TWELVE

The 'democratical' principle

IN JUNE 1830 GEORGE IV died after a long, painful physical decline, not much lamented by his country. His brother William, Duke of Clarence, became a delighted new king. Born in the year that HMS *Victory* was built, the old sailor was now 65 and determined to enjoy what years he had left. But the waves of political crisis that were about to engulf his reign were already lapping at the feet of the British establishment. The interests of industrial manufacturers, whose role in the British economy was far greater than it had been before the French wars, were still largely unrepresented in Parliament. Dissenters, many of them entrepreneurs, had for long been constitutionally barred from public office and now sought to test new liberties won with the Repeal of the Test and Corporations Acts of 1828. Agitation for reform by an emergent labouring class and by sympathetic radical intellectuals was an ever-present threat. Even more serious, breaches now appeared in the fortress walls of the governing Tory party, which had held power, apart from a very brief hiatus, for nearly fifty years since the younger Pitt's first premiership of 1783.

William Huskisson, Member for Liverpool, was the shining light of a liberal group of free-trade Tories. His enlightened series of tariff reforms during the 1820s emancipated trade with European countries and began to ease mercantile restrictions across the Empire by modifying ancient protectionist Navigation Acts. The Liverpool merchants were suitably grateful but Huskisson's attempt to ease the effects of Corn Laws by replacing fixed grain prices with a sliding scale of import duties was contemptuously thrown out by the overwhelming landed interest in Parliament. With his subsequent resignation, Huskisson inadvertently became the leader of the pro-reform wing of a party whose peers had fatally compromised their claim to represent the interests of the whole nation.

The cry for reform came not only from industrial interests, the middle class and the working poor. The repeal of the Test Acts in 1828 removed the obligation on Nonconformists to swear an Anglican oath before serving in the civil service, local government and the law. A broad coalition of Quakers, Methodists and Unitarians, passionately devoted to education and the abolition of slavery and emboldened by this liberal measure, now sought parliamentary representation. The same year the leader of the Catholic Association, Daniel O'Connell, was elected Member of Parliament for County Clare.

The new Prime Minister, Wellington, faced a personal and political dilemma. While the liberal wing of the Tory party had long advocated some form of emancipation for Catholics, Wellington (an Irish Protestant) and his Home Secretary, Robert Peel, were its implacable opponents. They refused to allow O'Connell to take his seat in the House of Commons. There were riots in Ireland. The threat of Irish civil war loomed. Wellington had had his fill of such things in Spain and so in 1829, with great reluctance, the government allowed a Roman Catholic Emancipation Act to pass. They had already alienated one wing of their party by their resistance to liberal reform. Now they lost the other wing, the so-called Ultra-Tories, by surrendering what had once seemed an inalienable principle of Tory Anglicanism.

In June 1830 the last thing Wellington desired was a general election. But after the death of the king the unwritten British constitution demanded it. Hustings were held in July and August as the great ancient rusty wheel of the electoral process turned once again. Rotten Boroughs returned government placemen; dukes, earls and barons wined and dined their electors; borough interests bribed, coerced and shouted down their opponents, while the liberal press and the mob agitated for universal suffrage, annual parliaments and the secret ballot. The complexion of William IV's first parliament was little different from the old. But the election had taken place against a backdrop of renewed revolution in France and the question of reform now took on an expediency unanticipated before the old king's death. Wellington resolutely denied the need for reform and predicted that there was no appetite for it in the country. He was wrong.

In Paris in July 1830, Charles X was deposed in a bourgeois coup which replaced his reactionary regime with a constitutional monarchy under Louis-Philippe, son of the celebrated Philippe Égalité. The revolution spread to the Low Countries in August and with tricolours once again being worn at political meetings across Britain, the clock seemed suddenly to have been wound back to the early 1790s. Once more, fears of a popular revolution being exported across the Channel spread fear through the British establishment. Parliamentary reform now began to look like a rational defensive option.

Henry Brougham's election for the County of Yorkshire was the most significant single result of the 1830 election. It was a tangible sign of the dissenting northern middle class flexing its new muscles. It gave Brougham sufficient moral authority to re-emerge as a populist Whig leader after the disaster of the queen's trial. The Tories might once again fear his eloquence; his own party feared, and with good reason, his ambition.

The summer of 1830 saw the third poor harvest in a row. During the elections in July and August, even as revolutions were breaking out across the Channel, disenchanted rural labourers in Kent began to expel Irish harvesters whom they believed had taken their seasonal employment from them.[1] At the end of August threshing machines, always seen as a threat to employment in the southern counties, began to be destroyed. It was the start of what became known as the Swing Riots, a wave of incendiary attacks and riots across southern England lasting almost until the end of the year, in which the burning of hayricks was an iconic feature. The unrest led to 19 hangings, more than six hundred gaol terms and five hundred sentences of transportation. The use of arson as a weapon of protest, symbolic of revenge, destruction and cleansing, had been psychotically perfected by Jonathan Martin. Its deliberate employment to terrorize the government during the Swing uprising seemed now to fulfil Mary Wollstonecraft's dire prediction of the potential evils of the Promethean myth.

Against this background and with an uncertain future facing Wellington's new government, the opening on 15 September 1830 of the Liverpool and Manchester Railway was an event almost impossibly overloaded with symbolism. Just as the analytical power of science, in the hands of Humphry Davy, had raised the spectre of Frankenstein's monstrous act of hubris, so this conquest of nature by the skill and graft of engineers and navvies seemed to offer limitless potential to tame nature on an even grander scale.

The boldest visionaries could see railway lines snaking almost uninterrupted from Calais to Vladivostok, from Dover to Edinburgh, from the east coast of America to the Pacific Ocean. September 15th was the day when the theory of mass transport would be proven or fall. There were, naturally, reactionaries for whom the prospect was horrifying: those, like Carlyle, who despised the Mechanical Age. An unnamed banker wrote to Charles Babbage, a keen student of the railway concept: 'Ah, I don't approve of this new mode of travelling. It will enable our clerks to plunder us, and then be off to Liverpool on their way to America at the rate of *twenty* miles an hour.'[2] This was not merely the first day of a new railway: it was the first day of the Railway Age. The completion of the line and rapid technical development of the Stephensons' new locomotives was a breathtaking achievement, a triumph of the Promethean spirit combining the empirical, practical bloody-minded determination of the father and the cool, professional scientific management of the son. Linking Liverpool and Manchester, those two great mercantile and manufacturing cities, made the Stockton to Darlington line, completed just five years earlier, look like what in reality it was: the parochial bastard offspring of eighteenth-century rustic improvisation. In the last year, since *Rocket*'s victory at the Rainhill trials, Robert Stephenson had transformed the locomotive into something instantly recognizable to steam engineers as in all essentials a fully modern machine.

That this might represent the birth of a great new era in communications and the transportation of goods was perfectly clear to the many thousands who turned up to witness the event. Among curious visitors were James Nasmyth, subsequently a giant of nineteenth-century engineering; Charles Babbage, taking a break from the fraught construction of his Difference Engine in London; and Fanny Kemble, the actress who had already ridden on the line and declared herself 'horribly in love' with the gruff Northumbrian George Stephenson.[3]

The directors of the railway were sensitive both to the financial importance of a successful opening and to its political overtones. The Duke of Wellington, as Prime Minister, and Home Secretary Sir Robert Peel headed a large cast of VIPs. The hero of Talavera and Waterloo, whose personal interest in technology and whose office demanded his attendance, was chief guest. But he had managed to make himself the most unpopular man in the country with his consistent denial of the merits of reform. His arrival in Manchester, a city still seething from the Peterloo Massacre of 1819 and burning with democratic ardour, might provoke riots. There was also the presence of William Huskisson to consider. The Member of Parliament for Liverpool was entitled to a place of honour in one of the several trains which had to convey hundreds of guests from one city to the other. But his recent resignation from the government, seen as a betrayal of party principle by Wellington, now saw him head a rival liberal faction in the party. The city he represented flew the banner of free trade. Could the directors of the railway contrive a reconciliation between the two men and with a grand gesture unite commercial, party and popular interests in a patriotic cause? It was an extraordinary risk to take.

The directors might just have pulled it off if, after the celebratory departure from Liverpool, Mr Huskisson had not been so slow to get out of the way of *Rocket* as he stood on the line at a watering station that fateful day. Every small child from then until now knows that to play on a railway line is a very dangerous thing. Mr Huskisson, the first man to be run over by a locomotive, did not, until it was too late. Fanny Kemble, fully sensitive to the sublimity of the occasion, wrote:

> After this disastrous event the day became overcast, and as we
> neared Manchester the sky grew cloudy and dark, and it began to
> rain. The vast concourse of people who had assembled to witness
> the triumphant arrival of the successful travellers was of the lowest
> order of mechanics and artisans, among whom great distress and a
> dangerous spirit of discontent with the Government at the time
> prevailed. Groans and hisses greeted the carriage, full of influential
> personages, in which the Duke of Wellington sat. High above the
> grim and grimy crowd of scowling faces a loom had been erected,
> at which sat a tattered, starved-looking weaver, evidently set there as
> a representative man, to protest against this triumph of machinery,
> and the gain and glory which the wealthy Liverpool and Manchester
> men were likely to derive from it.[4]

Huskisson's tragi-comic death (his leg was virtually severed when
Rocket's wheels ran over it and he died of his injuries the same night),
the bitterly violent reception of the duke at Manchester and the
organizational chaos which ensued took much of the shine off a technically successful opening. The first day of the railway had been designed
to celebrate a sense of national achievement. The directors cherished the
notion that it might unite two great cities and reconcile two wings of
the Tory party – Wellington and Huskisson had, in fact, shaken hands
in a rather poignant moment during the festivities. It remained to be
seen whether or not the Railway Age, and the Tory party, could recover
from the blow.

Huskisson's death, far from emasculating the left wing of the Tory
party, seemed only to encourage calls for Wellington to submit to the
demand for reform. Again in November 1830 he refused and as rioters
stoned Apsley House, with its famous address Number 1, London,
Wellington resigned his office. Earl Grey became the first Whig Prime
Minister since Lord Grenville in 1806–7 and the two most critical
years in nineteenth-century British politics began.

Among many sideshows during the summer of 1830 some had
consequences unperceived at the time. One hapless victim of the
election was the member for Guildford in Surrey, George Norton: a

rather boorish, greedy, unprincipled Tory from a large and active political family. Having lost his seat he prevailed upon his wife to use her influence to procure him some sort of government sinecure. She (they had married in 1827 when she was 19) was Caroline Sheridan, granddaughter of the unimpeachably Whiggish playwright Richard Brinsley Sheridan. They were an ill-matched pair from the beginning. Their families despised each other as a matter of political and social principle.

Caroline was bright, socially gifted, witty, determined and highly desirable to almost all men – at least all those who were not scared of her mannish conversation, not-quite manners and quick temper. Her social heritage gave her access to men and women of rank and talent across the broad spectrum of liberal politics and she revelled in her own charms. Soon after their marriage Caroline began to complain of her husband's temper, of her boredom with him, of his family. She patronized him and it fed his insecurity. He tore her letters up. She wrote more, many more. He burned her writing materials. He tried and failed to prevent her from going out alone. Then he began to beat her.

At the time when George Norton lost his seat the couple were living with their first child Fletcher in a house on Storey's Gate, just to the west of Parliament Square in the inner ring of London's political big top. Young literary men like Edward Bulwer and William Harrison Ainsworth were attracted to Caroline's company and fell under her spell. Ainsworth helped her to publish her first major work, a romantic poem called *The Sorrows of Rosalie*, in 1829. The Irish poet Thomas Moore, Byron's executor and author of a much-praised poem *Lalla Rookh*, and the young actress Fanny Kemble were frequent visitors at the Nortons'.

Caroline had been introduced to many of these exciting personalities by Samuel Rogers, the banker-poet (he wrote a couplet a day, no more and no less) whose taste and conversation were as oil to the social wheels of London's great social circus. Rogers was an old family friend who had eased Richard Brinsley Sheridan's poverty in his last years. At his house on St James's Place the Nortons were introduced to the likes of Wellington and Peel, Walter Scott, Wordsworth and the young Tennyson. Here Caroline met and became friends with the widowed Mary Shelley. Here too she met the Martins, many of whose friends she shared.

Despite Caroline's reasonably handsome earnings from *Rosalie* and *The Undying One*, published in 1830, the loss of her husband's seat threatened to wreck the Nortons' precarious finances. His post had been that of Commissioner of Bankruptcy in a government department which by the end of 1830 was being run by none other than Henry Brougham, who would certainly offer the post to one of his Whig supporters. It was entirely natural that, regardless of his Tory politics, Caroline should use her Whig connections to find George a job.

In the autumn of 1830 she began to write letters to the most promising of them and to some who she did not even know, but all of whom knew something of her background. One of these letters was sent to William Lamb, Lord Melbourne. His response to this intriguing correspondent, whom he had never met, inadvertently initiated a series of events that led to the first women's rights legislation and almost resulted in the downfall of the government.

However unanswerable the case for parliamentary reform may seem now, in 1830 there were still men of principle, Wellington among them, who believed that the existing system, despite its flaws, produced governments which acted in the best interests of the nation. They cited the successful administrations of the war years. They believed secret ballots to be un-English and underhand, for why should a man hide his intentions if they were honourable? The system that saw government placemen elected to nomination boroughs provided for a strong legislature. The influence of landed interest ensured that the traditional wealth-producing agricultural economy was protected against foreign competition. Perhaps even more persuasive to the sitting members in Parliament was the idea that a democratic system would fill the House of Commons with demagogues. Those members who had been classically educated, that is almost all of them, might have cited Thucydides and the downfall of Athenian democracy as precedent.

Wellington told the House of Lords in November 1830, shortly before his resignation, that: 'He had never heard of any measure up to the present moment which could in any degree satisfy his mind that the state of representation could be improved ... the legislature and the system of representation possessed the full and entire confidence of the country.'[5] The question was, would some kind of reform avert the sort of upheaval which had recently broken out in Continental Europe, or make such events more likely? The debate, increasingly bitter on both sides, was fought in the press, in the salons and drawing rooms of the middle classes and in the political unions and Mechanics' Institutes of the industrial north.

One unexpected beneficiary of the revolution in the Low Countries was Prince Leopold, uncle and still potentially regent to Princess Victoria and patron of John Martin. After the death of Princess Charlotte in 1817 Leopold became a sort of freelance. His prospects of happiness and utility apparently shattered, he began to study shells and was pronounced deadly dull by society. In 1829 he wedded an actress, Caroline Bauer, though the marriage seems to have been unconsummated and she left him the following year. He remained in close touch with his niece in England, studied European politics, and was flattered by an offer of the throne of Greece, newly recognized as an independent state. He refused, hoping for something better to come along and with half an eye on the British regency.

Then, in the same month that Wellington's Government fell, a conference was held in London to consider the future of Belgium. Britain and France became surprising allies in an attempt to prevent Austria, Prussia and Russia from invading the Low Countries. During the conference Lord Palmerston (a Tory defector like Huskisson) whom Grey chose to be his new Foreign Minister, met that great survivor of French foreign affairs Charles Maurice de Talleyrand to discuss their views on a suitable candidate to grace an independent Belgian throne ...

Palmerston to Talleyrand: Let us try and find someone ... who, by marriage, might satisfy everyone.
Talleyrand to Palmerston: I consider that everyone means you and us.[6]

In the end the newly freed Belgian people were coerced into accepting Leopold as a compromise, a safe pair of hands. He was crowned first King of the Belgians in June 1831, having carefully weighed the odds of his becoming regent in Britain and deciding that Belgium was a safer bet. Before even his first year as king had come to a close his new country was invaded by the Dutch. Eventually forced to withdraw by the threat of French invasion, the Dutch lost Antwerp to Belgium in 1832 and finally Leopold was able to set about building a nation almost from scratch.

A man whose promise of happiness had been so cruelly torn from him, who had seen so much dispossession and tyranny across Europe, Leopold had become something of a cynic. He later told his niece, Princess Victoria, 'The human race is a sad creation, and I trust the other planets are better organized.'[7] Nevertheless, Leopold set about building a nation with the same careful professionalism with which he had courted his princess. As a suitor he had surprised. As king he proved equally adept.

One of the immediate priorities for Earl Grey's new cabinet, formed on 18 November 1830, was to consider the burning question: not reform, but Brougham. Brougham's oratorical skills were outstanding. He was passionate for reform. But he was also a loose cannon, liable to fire off without warning and perhaps in the wrong direction. In the new government, delicacy must outweigh passion. The cabinet took the view that he should be made Lord Chancellor and elevated to the House of Lords. The appointment would please the middle classes who saw him as their champion and please his followers among dissenters and workers. It would also, crucially, remove him from the Commons where the cabinet believed he might do more harm than good. Sensing a bridle on both his political freedom and his ambition to become Prime Minister, Brougham demurred. But wiser voices prevailed, appealing to his sense of duty to the party, and he accepted.

Henry Brougham's sting having been drawn, or at least reserved for party service, Grey's immediate priorities were to set up a committee to consider what shape a reform bill might take, and to deal with the mayhem being caused in the southern counties by Swing rioters and what appeared to be widespread political agitation for a popular uprising. Throughout the autumn William Cobbett, that self-appointed watchdog of England's oppressed masses, had been lecturing to enthusiastic audiences in precisely those areas where Swing agitation was at its worst. Cobbett saw the rioting and incendiarism as the fulfilment of his often-repeated rhetoric that 'millions would take vengeance on thousands'.[8] The 11 December 1830 edition of his *Political Register* looked like an apology for violence and arson, made all the worse because Cobbett argued that the unrest had succeeded in forcing a lowering of much-hated tithes. If Cobbett saw himself as a more reasonable English version of Jean-Paul Marat, the new Whig Government saw the behaviour of this *ami du peuple* as a direct threat to their recent and precarious authority.

The new Home Secretary was William Lamb, now Lord Melbourne. Like many a liberal on taking up this post, he felt constrained to demonstrate his illiberality. In contrast to the light-handed way in which many local magistrates had dealt with Swing suspects, Melbourne set about putting the unrest down with extreme severity. Large numbers of sentences for imprisonment, death and transportation left no one in any doubt that he meant business. He decided also to prosecute Cobbett for sedition and incitement (in July 1831 Cobbett defended himself and was acquitted, like so many other targets of government vindictiveness in this period).

It was during his first few weeks in office, in the most fraught period of the Swing troubles, that Melbourne received the letter which Caroline Norton had sent him begging for a government post for her Tory husband.[9] Melbourne liked the company of interesting and beautiful women, despite or perhaps because of the experience of his marriage to Lady Caroline Ponsonby. Since the Nortons' home was close to the House and more or less on his route home, he decided to call on Caroline personally one evening. He must already have known something of her reputation.

They talked of the Sheridan legacy, of reform, of mutual friends. Caroline flirted sufficiently for her purposes, and very soon George Norton was appointed as a police-court magistrate on a thousand pounds a year, the salary of an admiral.[10] Caroline's conversation was erudite, witty, sometimes close to the edge of gentility, but always scintillating. Melbourne was urbane and worldly and he lived at the centre of political power in the world's most powerful country. There was an immediate, if platonic, mutual attraction. Melbourne became a frequent visitor and Caroline was often seen to wave at him from her window as he passed on his way to the Home Office every morning. The newspapers' gossip columns were delighted.

Caroline's drawing room, having already attracted most of the bright young things of London's literary set, now began to attract its aspiring politicians. On one occasion the young Benjamin Disraeli, then known if at all only as a novelist,[11] met Melbourne at the Nortons'. Melbourne asked him, rather patronizingly, what he wanted to be. 'Prime Minister', came the reply.[12]

More stable houses than the Nortons' were torn apart during the Reform Bill crisis of the next two years. George Norton may have bitterly resented his wife's politics and her universal allure, but he owed his job to her and their comforts to her handsome earnings as a poetess and now playwright and magazine editor. Somehow he managed to coexist with both his wife and his conscience. Given the mutual antipathy between the two families it is hardly surprising that the marriage continued tense and without love.

At the Martins', whose 'evenings' were among the most celebrated of the day for their eclecticism, music and sense of fun, the reform debate was carried on with great passion. John and Susan were staunch supporters of reform but they did not choose their friends solely for their political views. At Allsop Terrace a Tory bishop might sit down to whist with the editor of *The Examiner* (now Albany Fonblanque, whose wife's salute to John Martin in the park had so appalled Edwin Landseer, another occasional guest).

There is little sign that the Martins segregated their invitations; rather, in fact, they, and especially John, enjoyed the fiery spirit of debate which might erupt during an otherwise charming evening of laughter, poetry and song. The poet Tom Moore was a favourite guest, often to be seen at the Martins' piano, as were the son of Robert Burns,

James Hogg the Ettrick Shepherd and Thomas Campbell, another popular poet. Caroline Norton herself appears in Leopold Martin's lists of guests, along with Jane Webb, author of *The Mummy*, and many other young women writers whose careers were fostered by Susan Martin. Add to this mix the wonderful sight and sound of Charles Wheatstone and the more reserved but no less interesting Michael Faraday and it is no wonder that many a memoir of the late nineteenth century looked backed fondly on a Martin evening – especially if William Martin happened to be in town. The desire to have been a fly on the wall at some of these gatherings is almost overwhelming. Eye-witness accounts summon up a wonderful sense of vibrancy, full of the spirit of Prometheanism but also with a sharp taste of the political and social tensions almost universally felt at the time.

There can have been no more poignant guest at these gatherings than William Godwin, now in his mid-seventies and soon to have his long-standing debts finally remitted by accepting from Earl Grey a ceremonial sinecure: the Yeoman Usher of the Exchequer, with a residence at New Palace Yard at the heart of Westminster.[13] That the most revolutionary Englishman of the late eighteenth century should be comforted in his last years by a government pension is almost too perfect to be true. By now the old anarchist was, according to Leopold Martin, 'short and stout, with a remarkably large and curiously developed double skull, nearly bald. The little hair remaining on the temples and at the back was perfectly silvery. Godwin's eyes were deep sunken, shrewd, keen and lively, and retained all the fire of youth.'[14] During one of many such evenings at Allsop Terrace, according to Charles Macfarlane, a soldier and author who was present, Godwin was to be found happily and quietly playing at whist. He and Martin were long-standing friends, who often went for walks around London's old Dissenting cemeteries. Leopold makes no mention of Godwin's daughter, but given Susan Martin's predilection for intelligent women writers it seems likely that Mary Shelley was also an occasional visitor. She and her father had long been reconciled.

Macfarlane was surprised to find Godwin a 'quiet, retiring, unpretentious old gentleman' and that in the feverish climate of the reform debate he had moderated or abandoned many of the more radical ideas for which he was so famous. When a young man asked him what his fixed opinions were, Godwin apparently replied, 'I have none. I left off my fixed opinions with my youth.'[15] One would like to know if he was smiling when he said it.

While Godwin played cards, John Martin was in another corner, arguing passionately about reform with Alaric Watts, the journalist and poet. Martin was adamant that he would insist on pledges from parliamentary candidates, a hot topic of debate among radicals. Opponents of Martin's view believed that pledges would prevent members being able to speak or vote as their conscience dictated. When Watts contradicted him, Martin rose to his feet, claiming that everything he said could be found in Godwin's *An Enquiry*, which, indeed, it could, as could the opposite in that subtle and rational thesis.[16] 'Here is Godwin!' he cried, 'who will bear me out.'

Macfarlane recalled that Godwin,

> who was just sitting down to his *parti carré*, said that he might forget but he did not think he had written anything of the sort; that, if he had done so, he must have committed a great mistake, and that the imposing of pledges would turn a member of parliament into a mere delegate. The little painter and engraver was taken aback but he had too much vanity and vivacity to hold his tongue. 'But Mr Godwin,' said he, 'you will admit that your *Political justice* was all for knocking down the aristocracy, and for throwing the whole power of the nation into the hands of the people.'
>
> 'If I ever said so,' said Godwin, 'I must have been under a mistake.'
>
> 'Mr Godwin,' rejoined the artist, now getting rather vexed, 'I am afraid you do not stick to your principles.' The old reformed revolutionist, who was taking up his cards and arranging his suits, said mildly and even meekly, 'Principles and opinions! Opinions and principles! Perplexing things! When I really know what or which I am to stick to, I will think about making up my mind. It

is very easy to stick when, like a mussel, one sticks to the side
of a rock, or a copper-bottomed ship; when one doesn't think.'

'But,' said Martin, 'we have had march of intellect, progress
of education, intellectual development, throwing off of prejudices,
and now the Nation, the People, thinks.' [17]

It is almost as if one had overheard, during the late 1940s, Pablo
Picasso arguing about Franco and appealing to an aged George Orwell,
reluctant to interrupt his game of Monopoly. It is perhaps the fortune
of the best-loved revolutionaries that they die before ease and comfort
quench their fire. But Godwin, at any rate, was still a match for Martin:

Old Godwin, beginning to lead in trumps and transparently
annoyed at the interruption, yet still as calm and cool as a
cucumber, said, 'I don't think that a whole *people* can think.'

'Then,' said Martin, 'you throw up the democratical principle?'
'Perhaps I do,' said Godwin making a trick.

Reform aside, John Martin and his Promethean friends were still busy
envisioning their brave new world and trying to work out how to construct it. With the death of George IV Martin saw new opportunities
to attract royal patronage. The new King of Belgium had conferred
upon his old friend a knighthood of the Order of Leopold and
appointed him a member of the Academies of Brussels and Antwerp.
King William IV asked that *Nineveh* be taken to Buckingham Palace
so that he might view this famous historical painting for himself. Martin
was required to attend the king in person, but his hopes of becoming
Their Majesties' official historical painter were dashed when the king
came in, shook his hand, muttered 'How pretty!' as he glanced at the
painting, and walked on. Nevertheless, John used this slightly spurious
association to claim royal patronage for his *Illustrations of the Bible*, a
series which began to appear in 1831 and which made John's engravings almost universally recognizable in the parlours of English homes
of every social hue.

John had by no means given up on his schemes to bring fresh water
to the capital and deal with its sewage problem, though for the time

being his ideas fell on deaf ears. Soon even less inventive minds than John Martin's were beginning to concentrate on matters of public health. The Thames was still an open sewer. Too many people were being buried in the overcrowded churchyards and crypts of the metropolis. And within the year a devastating epidemic of cholera had broken out in the Martins' native northeast, reaching London at the beginning of 1832.

Equally momentous for the future, more so perhaps than the fate of the Reform Bill itself, were the goings-on in 1831 in the workshops of Charles Wheatstone, Charles Babbage and Michael Faraday. Wheatstone, whose interests had broadened from the family business of musical instrument manufacture towards sound, light and electricity, was forming ideas which led to the construction of an electric telegraph in the second half of the decade. His thoughts on how sound might be transmitted using electricity were the subject of a paper published in the *Journal of the Royal Institution* in 1831. The previous year, his proposal to accurately measure the speed of electricity, crucial in determining the feasibility of an electric telegraph, had been read out in the Institution's famous lecture theatre by his friend Faraday.

Wheatstone was still a popular visitor at the Martins', often to be seen with one of the instruments he had invented: the concertina, the harmonica or the 'enchanted lyre'. No great politician, he nevertheless sympathized with the educational passions of the Prometheans and so long as he did not have to deliver formal speeches, was content to entertain all-comers with some new experiment or practical trick. According to Leopold Martin, he was 'short-sighted, and with wonderfully rapid utterance, yet seemingly quite unable to keep pace with an overflowing mind'.[18] After he became the first Professor of Experimental Philosophy at the new King's College, London, in 1834 Wheatstone frequently asked John Martin to accompany him on his experiments, trailing wires across the Thames to demonstrate some aspect or other of his telegraphic theories. His work on measuring the velocity of electricity might seem trivial in comparison with his later fame as the inventor of the electric telegraph, but its conceptual beauty and engineering practicalities nevertheless indicate why he was so highly regarded by his colleagues.

Many years before, experimenters across Europe had tried to measure the speed of electricity through a solid conductor by sending a current down a very long wire (up to several miles in length) to see if sparks produced at either end appeared simultaneously. The problem was that if the sparks appeared at the same time there were two possible solutions: either the difference was too quick to measure, or the fundamental properties of the electric fluid had been misunderstood.

Wheatstone's apparatus was designed both to resolve this question and obtain, if possible, a constant for the speed of the fluid. He had already developed the use of a revolving mirror for measuring infinitesimal delays with his experiments on light. Now he adapted it to synchronize with the appearance of sparks at either end of his half-mile-long wire (the wire was coiled so that it was all contained within one room). Knowing the speed of the mirror's rotation, Wheatstone reasoned that the value of any delay in the mirror observed for the second spark could be measured to within some tens of thousandths of a second, an ambitious aim. He also anticipated the problem of a truly simultaneous spark. If it was the case that electricity was not a physical fluid but a state of flux rapidly alternating along a conductor, its passage might be of any speed and the sparks would still appear at the same time. He therefore provided for a third spark to be produced at the centre of the circuit.

Using many hundreds of coils, the three sparks were produced within a few inches of each other so that the visible deflection in the rotating mirror could be repeatedly observed. It was a triumphant experiment. The sparks at each end appeared at exactly the same time; the third, at the centre, was delayed by a factor which allowed Wheatstone to calculate the speed of electricity through wire as something over two hundred thousand miles per second. In fact, this exaggerates the true speed by a margin which further experiments reduced over the next few years. Its practical application was to allow Wheatstone to propose a practical system for transmitting information carried by alternating current over huge distances. Thirty years later, ninety thousand miles of telegraph were operating across Britain and London could communicate directly with Paris with a transmission delay of less than a second.

A short distance from Wheatstone's premises on Conduit Street, Charles Babbage was supervising construction in Dorset Street of his Difference Engine, a technology of no less long-term importance than the telegraph. Unfortunately the project, funded generously by the government under the personal patronage of the Duke of Wellington, had been hampered firstly by the extraordinary complexity of the device and secondly by the personality of the engineer Babbage had chosen to carry out the work.

Joseph Clement was a difficult man. The son of a hand-loom weaver from Westmorland, he was born in 1779, the year after Humphry Davy and Henry Brougham. He learned his trade as chief draughtsman under Henry Maudslay and was then recommended to Babbage by Marc Brunel as perhaps the only man capable of making the precision tools and components necessary for Babbage's design: all cogs, wheels and barrels worked to minute tolerances: rather like Bramah's unpickable lock – only more so. In fact, ownership of the tools became as hotly disputed as Clements' wage demands in the farce that led to the abandonment of the project. Clement, believing that through Babbage he controlled a direct line to government funds, overplayed his hand. He delivered the first part of the Engine to Dorset Street at some time during 1831. Then he packed up the tools he had designed and constructed for the Engine, and held Babbage to ransom for their possession. After that, nervous Whigs pulled the financial plug and no further progress was made, to Babbage's extreme frustration.

The engine, insofar as it was complete, worked precisely as Babbage had hoped, printing the results of its calculations in elegant type and dazzling those who came to marvel at its shining, clicking brass perfection. This is the machine which stands preserved and still working at the Science Museum in London. It was not until the Museum's own successful project to build Babbage's second Difference Engine in the late 1980s that any further practical Babbage design was completed:

more than a hundred and forty years later. The outrageously complicated printer to accompany it was finished in 2000; it works perfectly.

While Babbage was applying his fertile mind to the engineering applications of mathematics, his friend Michael Faraday was still exploring new theoretical and experimental worlds revealed by nature's most intimate secrets. This is not to say that he had become an unworldly boffin. His talents as a lecturer meant that he was in constant demand as the foremost explicator of science in England. He delivered many of Charles Wheatstone's papers, gave discourses on Marc Brunel's block-making machinery, was a public supporter of his friend John Martin's water schemes for the capital; and in 1826 inaugurated the celebrated Christmas Lectures series at the Royal Institution that continue to this day. His interests ranged from 'crispations', the term he gave to the patterns created in solids and liquids by sound waves, to industrial pollution and optical glass.

In 1829 Faraday took up a post as Professor of Chemistry at the Woolwich Arsenal. These burdens upon his time meant that his and Sarah's social life became increasingly restricted; they had no children, and their circle of friends was small and intimate. That it included poets, musicians and two of the foremost artists of the day, Turner and Martin, should not surprise. Those artificial barriers erected later by protectionists across all the disciplines had not yet been conceived. The Royal Academy and the Royal Society still occupied the same building, literally and figuratively.

Somehow, in 1831, with Britain apparently on the brink of revolution, Faraday found the time to make perhaps the greatest scientific discovery of the century. In retrospect, all his work since the dispute with Davy and Wollaston over electromagnetism in 1821 might seem to have been a distraction from pursuing this promising line of investigation. But Faraday was now 10 years older, a man of 40 and of wide reputation, his experimental curiosity spanning the whole of what we now call chemistry and physics, and beyond. The cumulative body of knowledge and technique which he had acquired in 20 years of study and experiment was about to bear Promethean fruit.

Sometimes scientists work on a pure hunch, trusting to their innate belief in serendipity (a word coined by Horace Walpole in 1754 meaning 'happy chance'). At other times, they construct an experiment with almost complete certainty of where it will lead. The experiment made by Faraday in his laboratory at the Royal Institution in August 1831 was of so specific a nature that it is impossible to believe it was not the product of malice aforethought. It involved a solid ring of iron an inch or so thick and a little over a foot across, which he had forged in the laboratory furnace. He wound 60 feet of insulated wire around one side of the ring, and a similar length around the other. The first coil was connected to a battery, the second to a magnetic compass. When current was applied, electromagnetic induction caused the compass needle in the second coil to deflect, even though there could be no physical conduction between the two. The same effect was produced when the current was turned off. The deflection of the needle must be the result of an electric current flowing in the unconnected circuit by magnetic induction. It was one of those moments that makes a scientist's blood run cold, like standing at the top of a waterfall, looking down.

Faraday wrote to his old friend and patron Sir Richard Phillips, 'I am busy just now again on Electro-Magnetism and I think I have got hold of a good thing but can't say; it may be a weed instead of a fish that after all my labour I at last pull up.'[19] This was disingenuousness, not modesty. Faraday took his ring with him when he and Sarah departed for a holiday in Hastings shortly after the first experiment. Scientists' wives get used to this sort of thing. He spent his weeks by the sea coiling hundreds more feet of copper wire, preparing for the next stage of the experiment when he returned to London. By the end of September he was absolutely certain that he had been able to convert magnetism into electricity and vice versa. By the middle of October, after more exhaustive and impeccably noted trials, he had shown that a magnet need only be near a coil to induce a current; and that an alternating wave of current was produced by the motion of a magnet towards and away from it. On November 4th he was able to write down the basic principles of the production of alternating current electricity. He had not known before he began with his coil exactly where the experiment would lead. But he had known that it must lead to something wonderful.

Careful to avoid any repeat of the unpleasantness ten years before, Faraday wrote an account of his researches in a paper and deposited it with the Royal Society at Somerset House on 24 November 1831. Exhausted, he fled to Brighton with Sarah to recuperate, leaving his fellow Prometheans to develop the application which powered the twentieth century.

If Faraday had stayed in his laboratory for the whole of that year he might have emerged blinking from the bowels of the Royal Institution in December to find that the new king had been crowned; that a Factories Act had been passed limiting children's working hours; that cholera had arrived in the northeast seaports of England; that John Rennie's new London Bridge was open; that a bill to reform the electoral system of the kingdom had been introduced, passed by the Commons but rejected by their Lordships; and that there had been widespread riots in protest.

Faraday would also have emerged just in time to see a young naturalist, Charles Darwin, grandson of chief Lunar man Erasmus Darwin, setting off on a voyage of discovery aboard His Majesty's Ship *Beagle* (she was the first vessel to sail under Rennie's new bridge). Perhaps sensing his own place in history, Darwin carried with him an ample supply of Samuel Jones's Promethean matches.

The progress of the First Reform Bill in 1831 was, to say the least, fraught. Introduced by Lord John Russell in March after a drafting process conducted in secret, it was more radical than many had feared, less so than others had hoped. In publishing it the government aimed to dampen popular agitation by showing its intention to reallocate a large number of rotten and pocket boroughs to 'new' centres of population in the north of England and in Wales (a separate bill for Scotland increased the tiny electorate there tenfold). It hoped to placate landed interests, including those of the Whig aristocracy, by keeping the ballot open and by offering the counties nearly half the reallocated seats. It sought also to throw a scrap to the middle classes by imposing a ten-pound property (or tenancy) qualification on urban electors.[20] As for the labouring masses, they would still have to rely on their betters to represent them in Britain's legislature.

Henry Brougham, now Lord Chancellor, was by design excluded from proceedings in the Commons. But, ever the Promethean, he personally arranged for the bill to be printed and distributed to newspapers across the country in large numbers by express immediately upon its publication, hoping that overwhelming popular support would disarm any attempt to suppress it in the House. It worked: just. The bill's Second Reading in the House of Commons was passed on 23 March 1831, on a night of the highest drama, by a single vote: 302–301.

This early promise was soon dissipated when, in its committee stage, the provisions of the bill were diluted to such an extent that the government abandoned it and dissolved Parliament. Prorogation was attended by great tension as suspicions of coups and Jacobin revolution had officials scurrying between king, Parliament and civil officials in near panic. On one occasion Earl Durham was said to have interrupted Lord Albemarle (Master of the King's Horse) at breakfast, demanding that the king's carriage be made ready so that His Majesty might be produced in Parliament immediately. Somewhat reluctant to forgo his meal, Albemarle replied, 'Lord bless me, is there revolution?' The earl responded, 'Not at this moment, but there will be if you stay to finish your breakfast.' [21]

In May another General Election was held. The country was in a feverish state and there was widespread intimidation of candidates and electors. Fortunately for the cause of reform King William, whose coronation was not due until September, proved himself a pliable if unenthusiastic supporter of the Reform Bill, for which he enjoyed huge popularity. The government's resignation gamble was repaid with a substantial Whig majority.

A Second Reform Bill was introduced in June. In July William Cobbett, defending himself skilfully against charges of sedition and incitement during the Swing uprising, was acquitted. It did not matter: Captain Swing had worn himself out. [22] With the prospect of real parliamentary reform the rick-burners and machine wreckers stood back and waited.

In September the Second Reform Bill passed in the Commons while in the north of England radicals discussed how to maintain pressure on Parliament without provoking desperate measures. Thomas Attwood, founder of the influential Birmingham Political Union in 1830, summed up the delicate position of the reformers: 'If we hold no meetings they say we are indifferent. If we hold small meetings they say we are insignificant. If we hold large meetings they say we wish to intimidate them.' [23] There were groups of extremists on both sides who wished the bill to fail. Conservatives hated the prospect of such a fundamental change in what they saw as a proud tradition of neo-feudalism. Many Radicals, like Henry Hunt, dismissed the bill with some justification as an anti-reform measure designed to protect in all essentials the existing system of political patronage.

Widespread cynicism towards the attitudes of Their Lordships in the Upper House was justified when, in October, the bill was defeated there largely by the intervention of a group of conservative bishops. Reformers' demands now hardened. In Bristol there were riots for three days. In Manchester, Leeds, Bolton, Derby and Nottingham large-scale meetings were called by the spreading Political Unions while powerful sections of the middle-class press demanded the formation of a National Guard to protect their property. Anti-reforming peers were stoned as their carriages passed protesters in the streets. In London the National Union of the Working Classes petitioned the government for universal male suffrage and annual parliaments, rather proving the moderates' point that the Reform Bill might be the least of several potential evils. Melbourne was forced to ban outdoor meetings of the sort which had so often led to riots in the past.

By the end of the year a new menace was abroad in England. The first victim of a cholera epidemic, which found public health provision in England's cities much to its liking, was a girl, the so-called 'blue girl' because of her symptoms, called Isabella Hazard, in Sunderland. Dr William Clanny, inventor of the miners' safety lamp, was one of the first to recognize the symptoms. More than five hundred people caught the disease there. Two hundred and five died. In nearby Newcastle, where a 19-year-old doctor called John Snow was serving his medical apprenticeship, more than three hundred died. Snow was the man who later crucially proved the waterborne cause of cholera during outbreaks in London in 1848–9 and 1854. By February 1832 the epidemic had arrived on a Newcastle collier in London (as it reached every other town in England, crossing the Atlantic to strike New York before the year was out). Among the victims in the capital was William Godwin's son, William junior. Across the country thirty-two thousand people died.

The government's response to crisis was to raise the reform stakes once again, threatening (to the king's intense annoyance) the creation of sufficient peers to force the bill through the Lords. When Grey calculated the number of new peers he might require, perhaps twenty, perhaps fifty, the king indicated that he might refuse. Tension in the capital was unbearable. Who would flinch first? The king, Earl Grey's inexperienced Whig ministers, the Radicals, or Their Lordships?

After making small, carefully calculated amendments to the provisions of the bill, and realizing that once again time was of the essence, Grey introduced a Third Reform Bill on 12 December 1831. This time it passed its Second Reading in the Commons by 324–162. Now for the Lords. Grey, in a delicate and expert series of manoeuvres, managed to persuade the king that the more enthusiastic he (the king) showed himself in support of the bill, the more likely that moderate peers would side with the government and the fewer the number of new peers he (the king) would be forced to create to ensure its passage. It was a brilliantly played hand. And it looked as though it would work.

On 14 April 1832, as cholera raged in London, the Third Reform Bill passed its Second Reading in the Lords with a majority of nine. But then, on May 7th, in the face of jubilant popular celebrations, the Opposition managed to carry a postponement vote. It was a disaster. Grey resigned and the whole cabinet with him. Britain was without government and close to civil war. The king had little option but to ask the most unpopular man in the country, the Duke of Wellington, to form a new government.

Now, London's City interests intervened, appalled at the idea of continued instability and, perhaps, revolution. How could Britain's merchants, industrialists and bankers operate in such a climate of uncertainty? They threatened to force a run on the Bank of England. On May 15th Wellington capitulated. Grey returned, and in June 1832 the Third Reform Bill became law.

In 1831 435,000 men, among a population of fourteen million in England and Wales, had elected 658 Members to the House of Commons. In the first reformed election of 1832, 650,000 electors were represented by an unchanged number of Members. It was true that seats had been redistributed. Many rotten and pocket boroughs had lost their Members while the larger towns and new industrial cities now returned one or two Members. The counties returned more Members than formerly and they were distributed more fairly. Any man living in a borough with a house worth more than ten pounds now had a vote. But there was no universal male suffrage. Not a single woman in the land was entitled to cast her vote. The ballot remained open, subject to bribery and coercion, and the parliamentary term remained at seven years. If the success of the Reform Bill is to be measured by the franchise, it was a thin sort of triumph for the democratical principle.

CHAPTER THIRTEEN

Babylon-on-Thames

IN THE YEARS AFTER THE PASSING of the Reform Bill
foreign visitors to London were in no doubt of its pre-eminence in
the modern world. Arriving in the city by ship from the Continent
one immediately saw that 'London is the real capital of the world ...
alone is entitled to talk of being the world.'[1] Independent of the
vagaries of tide and wind, smoke-belching paddle-steamer passenger
ships chugged right into the city's heart: past the East India Docks at
the mouth of the River Lea; past the vast complex of the West India
Docks on the Isle of Dogs; past Wren's splendid Royal Naval College
at Greenwich, then north around the long sweeping bow of Lime-
house Reach. At Wapping, just below the Pool of London, passengers
might see curious scenes of intense activity on the river, where Marc
Brunel and his engineers probed the riverbed with their diving bell,
still hopeful of healing the breach in the roof of the Thames Tunnel
after its abandonment in 1827.

Below Rennie's new London Bridge, within sight of St Paul's
Cathedral and the Tower, passengers were disgorged at a point where
land and river could scarcely be distinguished beneath a forest of ships,
boats and cranes: a ceaseless activity providing for the needs and wants
of a million and a half souls. From all over the world merchants, sailors,
tourists and representatives of all the known professions spilled onto
London's quays and bonded warehouses, agents' offices, dockside

taverns and waiting cabs. How could they all fit into London's already crowded streets?

One German tourist's overwhelming impression was of the value placed on time by London's rushing citizens. Another paused on Waterloo Bridge long enough to be intrigued by the novelty of iron turnstiles placed at either end to extract a penny from each pedestrian. Ninety thousand people walked to and fro across London Bridge every day. Away from the river one might effortlessly become lost in the maze of sooty streets, dazzled after dark by a profusion of gaslights or startled by the number of beggars and prostitutes. Even in the recently remodelled and more expansive sweeping boulevards designed by John Nash there was a permanent risk of being run down by one of thousands of carriages of every conceivable design (including Henry Brougham's eponymous four-wheeler whose rakishly elegant lines survived into the era of the Cadillac and the Rolls-Royce) almost choking the main thoroughfares. Pedestrians could take refuge in a sixpenny omnibus ride to the zoological gardens in The Regent's Park or any number of other attractions although, then as now, it might be quicker to walk.

There were more shops here than in any other city in the world. Not as elegant as those of Paris, Berlin or Rome perhaps, they nevertheless overwhelmed newcomers by their sheer profusion and the range of goods which they sold: the newest implements in agriculture and the mechanical arts; ironmongery, lamps in all their variety; prints, china, glass and porcelain. Curiosities from India and the Far East, from the islands of the Pacific, from Russia and Africa could be had at exotic prices. In the windows of druggists' shops large globes of glass filled with green, red and blue liquid were lit as if from within. Magically visible from far away, they served as lurid beacons advertising innumerable medical preparations available inside.[2] London was a vast city of business, impressive not for its beauty but for its air of impatience, its sense of an uncontrollable force riding pell-mell to its own doom (a doom predicted with relish by Jonathan Martin, for one).

Nothing contributed more to this tangible atmosphere than the capital's famously unhealthy fogs. Frederick von Raumer, the Prussian historian and minister of finance who visited London in 1835, wrote home:

> The root of most of the miseries is the London climate; – such, at least, as it has exhibited itself to me from my arrival up to the present day. It is true I see the sun, but not in his golden radiance; for though here is wealth enough to gild everything else, he alone appears red as a copper kreuzer, or pale as a silver groschen ... The thick fog which generally prevails is thoroughly impregnated with water, and this, blended with the air, is chilling and penetrative.[3]

Other visitors, especially those with artistic sensibilities, appreciated the particular quality of light produced by the combination of fog and the smoke of a million fires. Turner, from his shabby second home on the Thames at Chelsea, and later Whistler and Monet, celebrated it in paint. Ralph Waldo Emerson, the American poet, essayist and orator, visiting London the year after the passing of the Reform Bill and again some years later, caught the atmosphere of Old Testament doom perfectly:

> The smoke of London, through which the sun rarely penetrates, gives a dusky magnificence to these immense piles of building in the West part of the City, which makes my walking rather dream-like. Martin's pictures of Babylon & C. are faithful copies of the West part of London, light, darkness, architecture and all.[4]

John Martin's natural apocalyptic sensibilities chimed wonderfully well with both the fog and the prevailing mood of the new Reformed Parliament. By far the majority of his professional time was now devoted to mezzotint illustrations and to the unprofitable publication of his plans to improve the city's water supply, but he also found time to rework the subjects of some of his earlier canvases. In the lean financial period following his brother's arson trial, John hoped they might appeal to a public still reeling from the prospect of civil war. There was, for example, a new watercolour of 1833 on the subject of the *Seventh Plague of Egypt* and a much larger, much grander version of his *Deluge*, painted in the most foxy of hues a year later. This vision of God's wrath being wrought on his sinful creation is awesome and doom-laden almost beyond imagination. The sun is exactly the dim effusion of von Raumer's copper kreuzer. There is the trademark Martin bolt of retributive lightning shot from a terrible lowering sky. A sea of overwhelming immensity is about to sweep from the face of the earth a group of tiny huddling wretches in the foreground looking, as one rather insensitive critic wrote, like they had broken out of Bedlam.[5]

It was now more than a decade since the triumph of *Belshazzar's Feast*. John Martin was no longer the *enfant terrible* and such images, though they still evoked wonder, failed to shock. In 1833 John was elected a member of the Athenaeum Club under their famous Rule II, which allowed for the admission of eminent men not otherwise eligible. Here, a neat, handsome middle-aged and extremely well-connected artist of distinction, he was at home among the great men of letters and science. Contemporaries now wrote of him as an accepted part of London's social and artistic scene. He was a Knight of the Order of Leopold; he had been showered with gifts and honours by monarchs across Europe (though not in England). He wore his exclusion from the Royal Academy as a badge of distinction. As John Constable said, Martin 'looked at the Royal Academy from the Plains of Nineveh'.[6] Critics now routinely brushed aside the obvious defects in his technique. His technique was Martinian ...

> It would be worse than false to attempt to set up a standard of artistical comparison between John Martin and any other artist of the present day, or between his work and the work of any other artist. In this respect he resembles Sir Joshua Reynolds; he can be judged only by himself and not by comparison with another.[7]

John's friend Edward Bulwer-Lytton, whose own apocalyptic literary output included *The Last Days of Pompeii*, went much further with his praise in *England and the English*. Martin was

> ... the greatest, the most lofty, the most permanent, the most

original genius of his age. I see in him ... the presence of a spirit
which is not of the world – the divine intoxication of a great
soul lapped in majestic and unearthly dreams ... I consider [*The
Deluge*] the most magnificent alliance of philosophy and art which
the history of painting can boast. Look again at the Fall of
Nineveh; observe how the pencil seems dipped in the various
fountains of light itself.[8]

This is pushing sycophancy and perhaps taste into dubious realms. In
the same survey of British art Lytton argued that Turner had now
'forsaken the beautiful and married the fantastic'; that 'he no longer
sympathises with Nature, he coquets with her'.[9] History has judged
him wrong in his appreciation of both artists; but these were by no
means uncommon opinions in the 1830s.

Meanwhile, far away from London on a sun- and rain-swept Salisbury Plain, Martin's and Turner's splashy plein-air friend John Constable was even now creating one of the greatest watercolours in British
art. His *Stonehenge* is a work of revolutionary honesty, of scientific
brilliance more in tune with the work of Michael Faraday and Charles
Babbage than with either Martin or Turner. Intensely moody and
romantic, its portrayal of time's revenge on a forgotten civilization is
nevertheless perfectly in sympathy with London's antediluvian mood.

In the end it was not water but fire which a vengeful God visited on
his sinning London flock; and in a nice Promethean irony it was Turner,
not Martin, who was on hand to paint the great day of His wrath. On
16 October 1834 an unattended furnace deep in the cellars beneath
the House of Lords overheated through the negligence of two men
who had been charged with destroying an ancient accumulation of
wooden division tallies. These were sticks used to count votes in the
lobbies of the two Houses, made redundant after the dissolution of
the Court of the Exchequer in 1826. The furnace, left to its own
infernal devices, became red-hot and set fire to wooden panelling in
the cellars. By the time the fire was noticed, flames suddenly bursting
out near an entrance to the parliamentary complex, it had gained an
irretrievable hold.

All the fire engines of London could not prevent negligence succeeding in doing to Parliament what Jonathan Martin's ranting malice
had failed to accomplish at York Minster. The Westminster complex,
with its centuries of architectural accretions, its maze of entrances,
yards, courts and annexes, proved impossible to defend against the
inadvertent servants of Prometheus. On Westminster Bridge and in
hundreds of boats on the river, enormous crowds gathered to stare in
awed silence as the debating chambers of the Lords and Commons
were swallowed whole; only the ancient splendour of Westminster
Hall itself was saved.

The new Prime Minister, Lord Melbourne, was among the throng
who watched transfixed, apparently emotionless but grimly apprecia-
tive of the omen. J. M. W. Turner sketched obsessively from the
opposite bank, from Westminster Bridge and Waterloo Bridge. Recently
returned from Venice where his increasingly impressionistic visualiza-
tions were rendered in hundreds of almost frantic sketches, Turner
alone was able to capture the essential experience of those watching
the fire. *The Burning of the Houses of Parliament* is an eyewitness testimony of the fall of Babylon painted by the artist who, more than any
other, saw himself as the chronicler of old worlds passing and new realities dawning. Turner's composition evokes the sight of a theatre
audience hypnotized by a dramatic production of stupendous magnificence. This conception, together with the sheer luminescence of
the blaze, rendering the pathetic gaslights on the bridge mere dots of
pale yellow, is a technical and moral achievement outside the scope of
any of his contemporaries.[10] Perhaps England's dead revolutionary
poet Percy Bysshe Shelley would have done the scene justice had he
lived, but of her painters it is fortunate that Turner was on hand. He
might just as easily have called the painting *Prometheus Unleashed*.

The Prime Minister of a very fragile government, by nature
pessimistic and inclined to fatalism, whose seat of power is physically
consumed within months of his taking office, might understandably
have nurtured a desire to retire into the country and leave the
cares of government behind, feeling that the gods were against him.
Melbourne, the most reluctant of prime ministers and no follower of

Prometheus, would in the late autumn of 1834 gladly have handed over the reins of power if there had been anyone there to hand them to. There was not. The shepherd of the Reform Bill, Earl Grey, had resigned and returned to the quiet, verdant pastures of Northumberland. Henry Brougham, more desperate than most to lead a government, could not be trusted either by the Whigs or by the king. Of the alternatives, Viscount Althorp's father had died and his elevation to the Lords now left the Commons without effective leadership. John Russell and Edward Stanley had resigned over the simmering Irish Question.[11]

Melbourne was continually being badgered, either by the Duke of Wellington or the king or both, over Ireland. He was also riding the storm of public protest which followed the sentencing of a small group of Dorset trade unionists to seven years' transportation for swearing secret oaths. He had ignored a petition containing a quarter of a million signatures and a crowd of over thirty thousand people marching past his window demanding a pardon for the men from Tolpuddle. Now, in the aftermath of the fire when he went to see the king at Brighton and found he was to be dismissed along with his government, he seemed not to care very much.

In January 1835 a General Election (the fourth in five years) was held, not for the first time amid Home Office fears of insurrection from the trades unions, newly organized under a Grand National Consolidated banner by Robert Owen. The fact that Owen, the philanthropic New Lanark mill owner and campaigner for the cooperative movement, was a friend of Henry Brougham made the spectre of a Jacobin revolution all the more awful. It was only too easy to believe that Brougham was plotting a coup. In such circumstances a decisive result either way would have settled Whitehall nerves. But it was not to be. Robert Peel's Tories won but failed to make enough gains in the election to form more than a desultory imitation of an administration. So in March 1835 Melbourne was once again called upon to become Prime Minister. He was annoyed. A lazy, genial if sometimes downbeat man, he found the exercise of power a distraction from more interesting matters. He hated missing the theatre for the sake of a debate. He would rather have spent his time reading or flirting with Caroline Norton, whom he tried to see every day either before or after the business of his office was done.

Now Melbourne tried and failed to forge a coalition with the Tories, perhaps unwisely using Caroline as a go-between. His majority meanwhile had to rely on support from the Radicals and from O'Connell's Irish members, an uncomfortable alliance. With few potential colleagues to trust in forming a new cabinet, Melbourne nevertheless denied both Brougham and Wellington the chance to serve under him. Wellington was so furious that he nearly challenged him to a duel. Brougham, morally shattered by the prospect of his political career ending, by turns pleaded with Melbourne and threatened him. Melbourne wrote to him unmoved:

> You domineered too much with other departments, you encroached
> upon the office of the Prime Minister, you worked, as I believe,
> with the press in a manner unbecoming the dignity of your station,
> and you formed political views of your own and pursued them by
> means which were unfair to your colleagues.[12]

That was that. Melbourne somehow survived in office for another six years, almost reluctantly carrying through a series of reforms that mark the founding of the modern British state. Henry Brougham was left to pursue his many other interests while his contemporaries, however admiring they were of his huge range of talents, felt for the most part relief that his dangerous political sting had finally been drawn.

Henry Brougham's legacy lies in his ambivalent role as the political standard-bearer of the Prometheans, but perhaps most enduringly in his talents as an educator. In the face of entrenched establishment opposition he succeeded in engineering a public education system envisioned by a small group of passionately devoted believers. Harnessed to the technological potential of the steam engine, the imagination of the toolmaker and the skills of a generation of authors of whom he was himself in the vanguard (he is said to have written fifty thousand papers during his life), the Learned Friend's march of intellect succeeded in the end by exploiting the new mass-medium of cheap print.

Even as Brougham's political career was being dashed in 1835, Frederick von Raumer, that observant German visitor to London, was paying close attention to the liberating forces unleashed by the press. Visiting the premises where the *Penny Magazine* (run by Charles Knight as the main organ of Brougham's Society for the Diffusion of Useful Knowledge) was produced, he described how 20 steam-presses installed there were each able to turn out a thousand sheets an hour:

> Revolving cylinders are covered with printer's ink, which they spread over a horizontal surface, with greater evenness than could be accomplished by the most careful hand-labour. The machine takes the sheet, passes it over the types ... prints it on one side, then turns the sheet in the most intelligent manner, prints the other side, and deposits it before the hands of a workman who has nothing to do but take it away. And all this goes on more rapidly than one can tell it.[13]

Raumer, an astute philosopher and businessman, was not unaware of the implications of cheap printing. As he knew from many conversations with the thinking men of London it had played a large part in bringing about reform in Great Britain. It was a weapon that government found increasingly difficult to control through the blunt instruments of Stamp Duty and the Sedition Acts, and one which the Chartists and anti-Corn Law protesters of the next decade skilfully exploited. Having described the presses at work Raumer wrote home, rather prophetically,

> If there *were* a force which could effectively obstruct this infinitely accelerated power of diffusing thought, or could direct its operations at will, this would involve the possibility, indeed the actual existence of, a tyranny such as is unknown in history. In comparison with this, the red ink of censors were but milk and water.[14]

Raumer was in a position, with letters of introduction from very well-connected men on the Continent, to gain access to circles which included men like Wellington, Peel, Samuel Rogers and many of the

Prometheans. He took an interest in Turner (too much a 'nebulist', he thought), almost certainly met John Martin, and made time to attend the lectures of the celebrated Michael Faraday. He descended into the murky depths of the Thames in Brunel's diving bell. He visited many of London's clubs and societies, including the Athenaeum, and was impressed by what he saw as their enthusiasm for action rather than mere words. He saw Babbage's Difference Engine and tried in vain to understand the inventor's explanation of its workings. He reserved his greatest praise for Henry Brougham's brainchild:

> Of all the Societies, that for the Diffusion of Useful Knowledge is undoubtedly the most important. It had its source in the very just notions, that the civilisation of a people by means of reading is possible, provided really useful books were written for the people; and that these books might be printed at a very cheap rate, provided the numbers sold were large enough.[15]

The SDUK, as it was known, had managed to coerce the right authors into writing its publications. Brougham himself contributed many, as did other co-founders: Samuel Rogers, James Mill (father of John Stuart Mill) Josiah Wedgwood Jr and Rowland Hill, one of the most gifted teachers of his generation. The Society sponsored Mary Somerville's landmark translation of Laplace, though in the end it proved too large a project and was published by John Murray.[16] John Martin was commissioned to produce mezzotints for its edition of the Old Testament. The *Penny Magazine*, when launched, sold a hundred and fifty thousand copies and led to the publication of the *Penny Cyclopaedia*, issued in volumes over many years.

Thanks to the Society, working men and women had access for the first time to books on philosophy and history, on farming, mechanics and especially on science. The Society published series of maps and a biographical dictionary and many of its works became the first standard textbooks in universities.

Reactionaries hated it. Coleridge believed knowledge would not be diffused but diluted. Thomas Love Peacock derided the Steam Intellect Society, as he called it, for educating thieves; and

Bulwer-Lytton thought *diffusion* of knowledge all very well, but could not see how the Society might *advance* knowledge.[17] To many members of the aristocracy, the mere idea of popularizing knowledge served only to debase the arts and sciences. For Brougham, despite deep bitterness at the cauterizing of his own political aspirations, there was the consolation that he had helped to create a new force in Britain. He told Parliament in 1829: 'Let the soldier be abroad, if he will; he can do nothing in this age. There is another personage abroad ... the schoolmaster is abroad; and I trust to him, armed with his primer, against the soldier in full military array.'[18] For all the political establishment's doom and gloom and the fury of a working class which knew it had been cheated of its political aspirations by the Reform Bill, there was a new energy abroad in the late 1830s, and not just among schoolmasters.

In 1834, a long-cherished ambition of the Radicals was finally realized when Parliament voted to abolish slavery in its colonies (the trade in slaves having been outlawed by Britain in 1807 and by the Congress of Vienna in 1815). All slaves under six years of age were to be freed immediately. Slaves over six years were to remain part slave and part labourer for four years and then be free. Slave owners were to be compensated by a twenty-million-pound government fund.[19] Morally righteous as the Act seems now, and as it did at the time, its economic effects were disastrous on the plantations of the West Indies and South Africa. The clanking ring of cast-off fetters being forged into ploughshares was in reality as illusory a Promethean promise as the Reform Bill. Emancipated slaves had now to take their place in the overpopulated world of poor relief.

The 1830s proved to be the decade in which laissez-faire economics and the Utilitarian morality of public welfare, as opposed to private compassion, first drew the battle-lines of an epic conflict that has not yet been resolved. It is hard to deny the moral force of the welfare movement and yet, after each encounter social crusaders, landowners and industrialists left the field strewn with proletarian casualties. The Factory Act of 1833, the first to be enforced by independent inspectors, prevented children under nine from working in mills and controlled the working hours of those below the age of 18; it did not say how working families could make up for the loss of their children's wages.

The Poor Law Amendment Act of 1834 introduced the 'workhouse test' to ensure that able-bodied paupers were treated with uniformity across the country; but its imposition of a much stricter eligibility rule is now looked upon as one of the worst abuses of political self-interest in modern British history.

The Municipal Corporations Act of 1835 imposed a uniform electoral system on town and city corporations across the country. A more democratic measure in many respects than the Reform Bill, it nevertheless failed to outline the responsibilities which a modern borough ought to undertake on behalf of its citizens. Local police forces were set up to protect middle-class interests while provision for clean water supplies, public medical care and urban sewers was for the present ignored.

While the Whig Government pondered its legislative programme of ersatz reform, Prometheans were taking matters into their own hands. Cheap mass-market printing was already beginning to liberate access to knowledge. Now, a far greater social transformation was about to be wrought by the physical emancipation of Britain's citizens. The means of their emancipation was the steam locomotive.

While his father, seemingly in vain, tried to revive the fortunes of the Thames Tunnel project, Isambard Kingdom Brunel had begun to establish an independent career as an engineer. In 1830 he won a competition to design a bridge over the River Severn at the Clifton Gorge. Although the bridge was not completed until 1864, five years after his death, its radical design established Isambard's independent reputation, particularly in the southwest of England.

Encouraged and supported by his father's well-connected friend Charles Babbage, a great enthusiast for railways, Isambard also began to conceive of a railway line linking the port of Bristol with London. It was a project far more ambitious than the Stephensons' Liverpool to Manchester line which ran for a mere 35 miles. Its economic justification was unarguable: it would link transatlantic trade routes directly with the capital for the first time, bypassing both the capricious winds and tides of the English Channel and a number of canals. The Stephensons' line had already shown that canals were rendered uncompetitive by a reliable rail link. Bristol merchants, their families, agents and merchandise would be able to travel to London in hours instead of days. Doomsayers said it could not be accomplished, just as they had said the Liverpool to Manchester line could not be built.

At first, the group of merchants backing the plan for a rail link to London determined to accept the lowest bid from the competing engineers. But like the Stephensons, the younger Brunel was not a man to be easily stopped. He wrote to the committee, arguing forcefully that they would succeed merely in appointing the least scrupulous projectors to the scheme. The committee took the point, and in 1833 appointed Isambard sole engineer for the line.[20] The younger Brunel spent the next four months on horseback for up to 20 hours every day, preparing a survey for the line in which his guiding principle was to keep it as straight and level as possible: tunnelling, bridging and embanking wherever necessary. It would not be cheap. The merchants, in response, raised three million pounds in capital and prepared a bill for Parliament which was intended to circumvent those problems which had so long delayed Stephenson's projects in the north.

The first bill was nevertheless defeated in the Lords in 1834 under the influence of powerful West Country landowners who resented any sort of intrusion onto their ancestral property. Other opponents of the line argued that the planned two-mile tunnel under Box Hill, near Bath, would lead to wholesale destruction of human life, crushed beneath millions of tons of rock. No sane passenger would take such a risk. Owners of turnpike roads and canals protested against the line on economic grounds. The Fellows and Provost of Eton College deplored the proximity of such an abomination to their hallowed playing fields. Even the Duke of Wellington, such an ardent supporter of technology, had turned against railways once he realized that they would allow the lower classes to move freely about the kingdom. In the end the distribution of judicious quantities of money over the next year and the articulacy of Brunel's case secured the bill's passage on its reintroduction in 1835; but it was a near-run thing.

Isambard was no gruff provincial. The second-generation engineers had absorbed empirical engineering know-how with their mothers' milk; their managerial and diplomatic skills were acquired by travel and by exposure to the great men of the day; perhaps also by the embarrassment of seeing their fathers rubbing opponents up the wrong way. They were thoroughly professional. It was no longer inconceivable for a professional engineer to be seen in the circles into which Isambard married the following year. His wife Mary's family mixed with attachés, university professors and with musicians of the stamp of Mendelssohn and Schumann.[21] Marc Brunel had ensured an excellent education for his son. He sent him to the Lycée Henri-Quatre in Paris and then arranged an apprenticeship for him in the workshop of Louis Breguet, the brilliant French horologist and instrument-maker who matched Henry Maudslay not only in skill, but in founding a generation of machine-toolmakers.

Such an undertaking as the Great Western Railway was not without the usual problems associated with grand engineering schemes. Isambard came close to the point of resignation several times. Contractors had to be dismissed, landowners placated or bought off; huge increases in the estimated cost must be swallowed by financiers. There were substantial technical difficulties too, not the least of which was Isambard's insistence on using a new type of broad-gauge locomotive which he commissioned the Stephensons to build. He ended up altering it himself to increase its originally dismal performance. His letters from this period leave one in no doubt of his skill and determination. In one he might employ the subtlest of arguments laced with flattery. In another he would give splenetic vent to his frustrations; but even then his language was skilful and stylish, as one poor subordinate found to his cost: 'Plain, gentlemanly language seems to have no effect upon you. I must try stronger language and stronger measures. You are a cursed, lazy, inattentive, apathetic vagabond, and if you continue to neglect my instructions, and to show such infernal laziness, I shall send you about your business.'[22]

The same year that the Great Western Railway Bill was passed Isambard's father at last secured the financial backing to resume work on his long-abandoned Thames Tunnel. The failure of the original, on top of so many precarious years of debt-ridden worry, had left its mark on Marc Brunel. Now in his late sixties, he found his work was to be constrained by Treasury conditions attached to the Tunnel's capital loan. He looked and felt old. But in March 1835 the bricked-up works were reopened, the old shield removed and a new, larger mechanism that had been designed by John Rennie (by now long dead) was installed.

All the old drainage and tunnelling problems returned to plague the enterprise: irruptions of filthy, poisonous water; noxious black mud, and firedamp. Tunnellers complained of headaches, nausea, and diar-rhoea; some of them died. The lack of provision for drains and venti-lation was not for want of care on Brunel's part. Applications to the Treasury for funds to install fans, pumps and channels for drawing off water and mud were turned down.

Despite these problems the tunnel progressed, yard by painful yard. For four years Marc was sent samples of excavated soil every two hours, day and night, via baskets attached to a pulley system from his bedroom window. Towards the end of the year 1840 he wrote to a friend,

> The four Elements were at one time in particular against us; *Fire* from the Explosive gases, the same that are fatal in mines; *Air* of a mephitic character, by the influence of which the men most exposed were sometimes removed quite senseless; *Earth* from the most terrific disruptions of the ground; *Water* from five irruptions of the river, three of which since the resumption of work in 1836! [23]

The unfathomable breach in the roof was finally plugged with the aid of a newly developed material: Joseph Aspdin's Portland cement. This was a form of concrete whose characteristics included its propensity to set under water. Engineers simply poured the mix into their riverbed breach from boats. The Tunnel finally opened to pedestrians in 1843, 18 years after digging began. It was converted to carry the East London Underground Line in 1869.

That this pioneering engineering project mirrored the history of coal mining was reinforced when, in the first year after the resumption of works, an explosion ripped through the Church Pit at Wallsend Colliery on the north bank of the River Tyne, the scene of so many attempts to conquer firedamp through ventilation. One hundred and two men and boys were killed. By some sort of miracle, a small number of men were brought out alive. No one knew what had caused the explosion, but there were suspicions that the pit's long record of safe working had led to the use of naked flames, or to some neglect in the use of their safety lamps.

Britain's legislature (still awaiting a permanent home after the fire of 1834) sat conveniently far from the coalfields of the North. No proper records were kept on the causes of accidents in mines, or for that matter on canal or railway construction. Parliamentary minds were generally only concentrated on such matters when a disaster was sufficiently close or grand to excite London's drawing rooms.

By coincidence the government had already appointed a Select Committee on Accidents in Mines to look at an industry whose operation was unregulated, in which working conditions were entirely a matter of industrial self-interest and private philanthropic concern. John Martin, whose elder brother William still boiled with indignation that his own, superior lamp had never been adopted, gave evidence to the Committee on three separate occasions. He clashed with none other than George Stephenson on their respective schemes to ventilate mines, reopening the old debate about safety lamps which so exercised William and which exercised John on his behalf.

If John Martin's contribution to the debate was a disinterested act of civic duty, he was in a minority. Under the influence of colliery owners, viewers and merchants, the Committee concluded with casual apathy:

> The Committee do not recommend any particular new plan for improving safety, but draw attention to the known measures which are not always implemented. They appeal to men of science to interest themselves in these issues. Greater attention should be paid to the education of miners, and the efforts of some colliers in this

respect are admirable. Mine-owners, however, should be reminded of the great responsibility they hold for the lives of their employees, and the importance of adequate numbers of competent managers and overseers should be stressed.[24]

For the first time, it was determined that local coroners should record the cause of death in mining accidents. But miners had to wait to be afforded the same legislative protection that mill workers were beginning to enjoy.

The first section of Isambard Brunel's Great Western Railway, from Paddington to Maidenhead, was officially opened in 1838. In the same year Robert Stephenson's 112-mile Birmingham to London railway was completed, with its terminus at the new Euston Station. The line's passage through London's northern suburbs required the demolition of thousands of homes. Twenty thousand people had to be rehoused. The middle classes of the metropolis responded by moving further out into more tranquil parts, establishing the first commuting towns along the new route. New lines were built to Greenwich and Blackwall.

No less than 44 railway companies were established during the following two years. Charles Pearson, Solicitor to the City of London, proposed the extraordinary idea of an underground railway running along the Fleet valley from Farringdon. It was not long before John Martin added a circular underground railway to his ever-evolving plans for the capital. Across the Channel in Belgium his friend, the enthusiastic technophile King Leopold, had already had the Continent's first trunk line constructed between Brussels and Mechelen.

Public response to the arrival of the railways was ambivalent. Passengers embraced them with an enthusiasm bordering on mania. Old-fashioned men deplored them. Employers feared them. J. M. W. Turner saw in them dramatic narrative and atmospheric possibilities. Tunnels, already explored as a metaphor by John Martin in his darker compositions, became a Victorian icon of the struggle between good and evil, between ambitious men and the elements. Nobody expressed this ambivalence better than William Wordsworth, former political zealot and worshipper of nature in her pristine virginity. Since Shelley, no writer so perfectly caught the moral dilemma of the Prometheans as Wordsworth achieved in the poem 'Steamboats, Viaducts and Railways' which he wrote during a summer tour in 1833:

> Motions and Means, on land and sea at war
> With old poetic feeling, not for this,
> Shall ye, by Poets even, be judged amiss!
> Nor shall your presence, howsoe'er it mar
> The loveliness of Nature, prove a bar
> To the Mind's gaining that prophetic sense
> Of future change, that point of vision, whence
> May be discovered what in soul ye are.
> In spite of all that beauty may disown
> In your harsh features, Nature doth embrace
> Her lawful offspring in Man's art; and Time,
> Pleased with your triumphs o'er his brother Space,
> Accepts from your bold hands the proffered crown
> Of hope, and smiles on you with cheer sublime.[25]

When, a decade later, his own beloved Westmorland was threatened by the construction of a railway line, Wordsworth took a rather different view in a poem called 'On the Projected Kendal and Windermere Railway' – an early example of what is now termed 'Nimbyism'.

Most of the Prometheans were resolutely enthusiastic towards the new mode of travel. Charles Babbage, who had attended the opening of the Liverpool and Manchester Railway in 1830, invented a 'pilot' attachment for the front of locomotives to keep lines clear of anything from snow to George Stephenson's 'varry aakward coos'. He was a staunch supporter of Brunel's broad-gauge tracks and made speeches in their defence. He also invented a device for measuring the speed of a locomotive, though it was not taken up with any enthusiasm.

Charles Wheatstone began to see that railway lines might be used as arteries to lay lines for the telegraph system that he was now working seriously towards. In 1841 Isambard Brunel invited Wheatstone and John Martin to accompany him on a test run in one of his broad-gauge locomotives. The experiment: to find out how fast it was possible to travel on a locomotive without, as some predicted, stretching the fragility of the human body beyond its limits. They met at Paddington and, accompanied by timing-clerks and under the supervision of an experienced driver, took the locomotive a few miles out to where a substantial section of line had been cleared of all traffic between two timing points. No carriages were attached. John, as was his practice on most of his expeditions, had brought with him young Leopold (not now so young: he was 24 and had been given a job in the Stationery Office by Lord Melbourne):

> The start took place from Southall, the engine being provided with a thick plate glass screen in front to protect those upon it from the strong current of air. With Mr. Brunel's eye on the gauge and hand on the safety valve, we were off, and in half a mile we were running at 'top speed', the time-keepers busily at work. To the great satisfaction of Mr. Brunel and the astonishment of all, it was discovered that the distance of nine miles from the station at Slough had been run in six minutes, or at the rate of ninety miles an hour – a very different result from that which Mr. Stephenson's early calculations would have led one to expect.[26]

Despite worries over his declining financial situation, John Martin remained as sociable as ever. His old friend William Godwin had finally died in his 78th year, much lamented by John; but there were new acquaintances to be made, sometimes in surprising places. On one lengthy sketching holiday in East Anglia, John and Leopold visited a number of friends: an engineer who was building a road across the fens; Sir Francis Chantrey the sculptor, and Dean Peacock, busy restoring the glories of Ely Cathedral. They went to King's Lynn and Hunstanton, and at Castle Acre Priory they were shown round the ruins by a most impressive lady dressed in the costume of the Quakers. When, after enjoying an interesting and informative tour, they swapped cards, the lady's proved to be that of Elizabeth Fry, the celebrated social reformer.

Fry was in her mid-fifties. Not content with bearing and raising

11, perhaps 12 children, she had revolutionized the treatment of women prisoners by her work at Newgate in London. She was the first woman, in 1818, to be summoned before a Parliamentary Select Committee, to give evidence on prisons. She campaigned to improve the lot of convicts transported to New South Wales and founded London's first night shelters for homeless beggars. By the time she met John Martin she was as famous as he was; they had admired each other's work for many years.

In 1836 John and Leopold were introduced by John's friend George Cruikshank (who to Leopold's delight used to draw faces on his fingernails or shirt cuffs when he was out without a sketchbook) to a young writer whose first novel, called *The Pickwick Papers*, was just then being published in monthly parts. Charles Dickens, 'full of fun and dash', had been married for only a short time. 'His fair hair waved long and freely over a white and unwrinkled forehead, and a clear and healthy complexion; his eyes were large, bright and penetrating and he had a smile of glee for all ... No care had ploughed its lines; yet the countenance had already acquired all the ideal of a great intellectual presence.' [27] Railways and Charles Dickens made their stamp on British society at the perfect moment to ensure their permanent association with the dawn of the Victorian era. On 24 May 1837, in the nick of time, King Leopold's niece Princess Victoria gained her majority. Within a month her paternal uncle King William IV was dead and the young, serious but perfectly engaging girl was queen. Showing few signs of nerves and with an air of self-possession which astonished those who met her for the first time, she conscientiously went about her duties under the tutelage of her first Prime Minister, the otherwise reluctant Lord Melbourne.

They fell into immediate habits of friendship. After the gruffly bullish King William, Melbourne found Victoria refreshingly sensible, perceptive and consistent. In him the queen found a man whose disinterested concern for her was a rock of trust in a confusing world of social unrest, petitions, party squabbling and self-serving patronage.

The first years of the new reign might have been very different if, in the year before Victoria's accession, King William had accepted

Melbourne's offer to resign his premiership. He had been cited as co-respondent in a divorce case which was one of the most scandalous of the age. The other parties were George Norton and his wife Caroline.

Their marriage had barely survived the Reform Bill crisis. Norton's family were old Tories: opponents of reform, of Catholic emancipation, of the abolition of slavery, of anything smacking remotely of Whiggery. Caroline's impeccable liberal credentials were inherited via her parents from her grandfather, the playwright and MP Richard Brinsley Sheridan. She had an atavistic urge to dip her toe in the muddy waters of Westminster, the more so since her friendship with Melbourne. In 1832 she wrote impassioned letters to potential Whig supporters, urging them to become candidates for the reformed House of Commons. Charles Babbage, receiving one of them, went so far as to stand in the general election, though he did not win his seat.[28] In so openly playing politics Caroline made herself an easy target for the wrath of the Nortons. When the marriage failed, as it must, she found them more implacable even than her boorish husband.

After one quarrel in 1833, during which Caroline locked George in the dining room, he took his revenge by destroying her bedroom and throwing her down the stairs – this within a few weeks of the birth of their third child. The following year, during a Sheridan family trip to the Continent, George fell ill. Caroline stayed with him while the rest of the party moved on. There was another violent quarrel, probably about the cigars he smoked which Caroline so hated. This time the Sheridans failed to hide their contempt for Norton. They persuaded Caroline to leave him behind while they moved on to Paris and here, quite naturally, the three Sheridan sisters caused a sensation with their wit and dashing beauty. Henry Brougham was there at the time and paid a great deal of attention to Caroline. But members of the Norton family were also in Paris. Caroline's apparent abandonment of her husband and the Sheridans' unswerving support for her side of the dispute turned a marital rift into a family feud whose chief victims were Caroline's children.

In 1835, when Melbourne tentatively offered peace and a coalition partnership to the Tories, Caroline's part in the proceedings became known to some of those excluded from the discussions, including the Duke of Cumberland and Henry Brougham. By association with Melbourne she was damned in their eyes. Caroline had inadvertently made two powerful enemies. That autumn, perhaps as a result, a series of scurrilous articles appeared in popular sections of the press insinuating an affair between Caroline and Melbourne. In truth they had not been discreet about their friendship. Melbourne had a long-standing amorous reputation and Caroline's public flirting was thought unbecoming by a suddenly prudish London set.

When the final separation came after yet more violent quarrels, it provided an irresistible opportunity: for George to extort money from a suitable co-respondent; for his family to avenge themselves on the Sheridans by taking Caroline's children from her; and for a group of opportunistic and thoroughly disreputable Tories to unseat the Prime Minister.

George's first thought seems to have been to name the young novelist William Harrison Ainsworth in divorce proceedings. Other names were proposed: Shelley's old friend Edward Trelawney, who had found the poet drowned on the beach at Viareggio; and the Duke of Devonshire. Caroline had so many admirers and was so open in enjoying their admiration, that it came down to a question of which name on the suit would prove most advantageous. In finally naming Melbourne, George and his self-interested advisers raised the stakes very high indeed. The thought of bringing down the government must have seemed superficially a very attractive idea to the Nortons. A new Tory administration could hardly fail to reward them for such a dramatic coup. If the case succeeded Norton might make himself rich at Melbourne's expense and the children might be protected from the influence of an unsuitable mother. On the other hand, George had compromised himself by actively encouraging his wife's friendship with Melbourne and by himself begging the Prime Minister for money. It was a dangerous game in which there might be no winners. If George Norton had been more intelligent, or at least less suggestible, the suit might have been abandoned. But it went ahead.

A flurry of letters passed between Caroline and Melbourne, whose visits to her in public and in private abruptly stopped. He tried to reassure her of his support even though he saw his own position as untenable, regardless of the outcome of the case. He did not play the classical part of the shamed minister: he had nothing to be ashamed of and was indifferent to the loss of high office. His strong sense of affliction was on her account and on that of his political colleagues, not for himself. She, with a chasm of social ostracism opening up before her and unable to see her three boys Fletcher, Brinsley and William, was frantic. She almost instantly became a social pariah, a woman abandoned (indeed ejected from the family home) by a husband who believed himself to have been cuckolded.

With the impending divorce case seized upon by the press, Melbourne felt he had no option but to go to the king and offer his resignation. William refused it, suspecting the sort of covert Whitehall plot that no man brought up in the Royal Navy could countenance. Wellington, who had perhaps the most to gain from Melbourne's embarrassment, also offered his support: he would not serve in a government brought to power by such shabby means.

Neither Caroline nor Melbourne appeared in person when the case was heard in June 1836 under a confetti of salacious press articles and cartoons. George had procured a number of witnesses from among Caroline's former servants. He was able to produce some trivial correspondence between the two alleged lovers of the 'shall I see you today?' sort: very thin gruel for the press. Melbourne's counsel, in the impressive shape of the Attorney-General Sir John Campbell, made light work of Norton's case. His cross-examination of witnesses was masterly; his suggestion of a political plot perfectly convincing. There was, as Norton's friends might have advised him, no case to answer, for the two alleged lovers, for all their intimacy, had never had an affair – or if they had it was the unspoken, unconsummated affair of two mutually attracted friends. On his return to the House after the jury's rapid acquittal the Attorney-General was cheered by Members on both sides.

Melbourne survived to meet his queen the following year. Norton retired to lick his wounds and contemplate further revenge. Caroline had to busy herself with writing and with a long campaign to regain access to her boys. Perhaps naturally she identified the plight of her boys with that of the thousands of children who lived effectively as slaves of the manufactory system. In the year of the adultery trial she wrote a long poem, *A Voice from the Factories*, which pleaded with Parliament to legislate on the conditions of child labourers. It was accepted by John Murray, the publisher of Jane Austen, Mary Somerville and Lord Byron. It was not great poetry. Sentimental and polemical, it failed as a manifesto, but in its self-conscious phrasing it nevertheless reflects Caroline's natural Promethean sentiments:

> Beyond all sorrow which the wanderer knows,
> Is that these little pent-up wretches feel;
> Where the air thick and close and stagnant grows,
> And the low whirring of the incessant wheel
> Dizzies the head, and makes the senses reel:
> There, shut for ever from the gladdening sky,
> Vice premature and Care's corroding seal
> Stamp on each sallow cheek their hateful die,
> Line the smooth open brow, and sink the saddened eye.[29]

Now socially isolated, her independent income at the mercy of her husband's lawyers, Caroline fell back on a tight circle of friends and the battalions of indignant Sheridans. She had no option but to fight. It was a battle that she waged purely for herself but which eventually led to the first legal rights for all women.

CHAPTER FOURTEEN

Survivors

SHELLEY, KEATS, BYRON, AND JANE AUSTEN had all died too young. If the surviving Prometheans wanted to see their heavenly visions materialize on earth they must live long, if not necessarily prosper. The generation born between 1770 and 1790 were now mature and those with undimmed purpose and energy tapped into the energetic atmosphere of Victoria's early years.

John Martin produced much of his commercially successful work in his thirties but some of his most important paintings were produced during his last few years. J. M. W. Turner would have left an impressive legacy if he had died in 1830, but the works which confirm him as England's greatest painter belong to the two decades after that. Constable died in the year of Queen Victoria's accession, a year after exhibiting his watercolour masterpiece *Stonehenge* and still surely unfulfilled.

Michael Faraday discovered and developed electromagnetic induction in his 40th year; in his 49th he was jointly responsible for introducing photography to England. It must have seemed that he had completed his life's work when, in 1839, he began to suffer from a debilitating illness that lasted some five years. But he recovered sufficiently to prepare the ground for a number of the key developments of twentieth-century physics with his Field Theory.

Charles Babbage, a year younger than Faraday, had built his Difference Engine at a similar age and still had at least one great idea left in him. His Analytical Engines, a series of dazzling mechanical and mathematical conceptions which have only in the last few years begun to be physically realized, occupied him until his death at the age of 80.

Charles Wheatstone, a decade younger, was as yet barely famous. George Stephenson saw the opening of the Liverpool to Manchester railway in his 40th year, while Marc Brunel was working towards his final triumph well into his sixties. Elizabeth Fry embarked upon her last great project, the establishment of a school of nursing, at about the same age. Mary Somerville's final important work *Molecular and Microscopic Science* was published in 1869 when she was almost 90. In 1834 she had been the first person to whom the term 'scientist' was applied, by William Whewell. Henry Brougham may have committed political suicide in his early fifties but his educational and academic work continued until his death.

Brougham's exact contemporary Humphry Davy was the exception, having achieved his greatest triumphs in his twenties and early thirties and having burnt himself out ten years before his death in 1829. Others, like Mary Shelley and Leigh Hunt, were similarly meteoric, their reflected glory fading with the death of their star Percy Shelley while the radical poets of the revolution, Wordsworth, Southey and Coleridge, had lost either their fire or their purpose.

In Victoria's first years, then, the Prometheans still had much to offer. The new queen's youth, charm and purity gave a tired country (and a tired Prime Minister) new energy. Indeed, it was the queen's coronation that ended the long financial and mental decline which John Martin's obsession with clean water and sewage had brought about. He decided to paint it. In the year of Victoria's accession, 1837, Martin felt himself 'a ruined, crushed man'. 'I shall sink now,' he told his friend Ralph Thomas ... 'There are no more bright days for me. My eyes have been opened to the state of my affairs and I am a pauper. I am dishonoured. I know not what will become of us. I have never attended to money matters and this is the consequence ... I have been plundered and deceived.'[1] John's financial worries had begun with the failure of his bankers in 1824 and worsened with Jonathan's trial for arson in 1829. At that time he believed he would always have a regular income from engraving – he had every reason to believe so. But unscrupulous pirates in Britain and on the Continent stole his designs. He successfully sued at least one and argued alongside Turner before a Select Committee for a tightening of copyright laws; to no avail.

John Martin's income began to fall just as he turned his attention away from commercial art towards his schemes for improving the capital's water supply. So far these schemes had come to nothing. He had little idea of the relationship between income and expenditure. He often lent money to his brothers and friends and saw very little of it back. He had an ailing wife and a large brood of children to support, and he and Susan had adopted Jonathan's son Richard. The famous Martin evenings, enjoyed by London's elite over many years, had been a continual, if rewarding, strain on his resources.

At least one debtor was now beyond redemption. John's brother Richard died in 1837 at the age of 58. After his retirement from the army in 1827 he had been living in London, writing verse but not apparently making a living. His pension was eked out by frequent subventions from John, who always gave him money when he asked and always complained bitterly when the sum was not repaid.

John's first idea, on the new queen's accession, was to apply for the position of Historical Painter to Her Majesty, using his friend King Leopold, the queen's uncle, as an unspoken referee. But his claims went unanswered – he had not yet cultivated the valuable acquaintance of Leopold's nephew Albert of Saxe-Coburg-Gotha. By the beginning of 1838 Susan Martin was visiting Ralph Thomas with worrying accounts of John's physical and psychological state – evidently she was worried that he might finally betray the signs of mental ill-health that had brought Jonathan to destruction. He talked to no one. He looked haggard and sullen.

In May Jonathan died in Bedlam. His son Richard, who had lived with John's family on Allsop Terrace in Marylebone for nearly a decade, wrote to William in Newcastle with the news.

My Dear Uncle,
I do not write to you often, and when I do I have only to communicate occurrences of a melancholy nature. I last year wrote and

told you of your brother Richard's decease, and now I have to announce that my poor father is released from his bodily confinement: his spirit has escaped the boundaries of his prison walls, and returned to Him Who gave it and Whom on this earth he made it his duty diligently to serve …

My aunt Atkinson [Jonathan's sister Ann] and I were there on the Monday, two days before he was taken ill; he was delighted to see us, and appeared cheerful and contented: he talked about his friends in the north, and asked if we had heard from you, and hoped that you were doing well. He had got permission to draw a little, which he had not been allowed to do for a long time …

He did not keep to his bed until Saturday when he felt himself a great deal worse, especially towards the night. The keeper came and told us of the condition he was in. My Uncle [John] and I immediately went to see him, but he was in a state of lethargy and insensible to everything around. He did not know us …[2]

Jonathan's death brought more misery to the Martin household, but worse was to come. It is not clear whether it was the latent fragility in the Martin genes or a feeling of guilt that he was contributing to the family woes that induced Jonathan's son Richard, then 28 years old, to kill himself in August, three months after his father's death. This fresh tragedy might have sent John over the edge of his own mental precipice, but it did not. It seems in fact to have brought about a catharsis. He somehow cast off his depression and set about the business of making money again.

Ralph Thomas' account (or at least Mary Pendered's rendering) of the Martin revival may be a little rosy, but it is entirely in keeping with John's determination to use his own resources in overcoming the fates that seemed to obstruct him at every step. He had been indicted by his first employer in Newcastle. He had lost his first proper job before it started and spent many years making a miserable living on the streets of London. His co-workers, jealous of his talents, had struck against him. He had been spurned by the Royal Academy and scorned by critics for three decades. His brothers were an embarrassment that more than once threatened his reputation. Adversity was nothing new.

The queen's coronation had already been painted by the distinguished artists Sir George Hayter and Edmund Parrish. But Martin was probably present at the event and knew that with his unmatched architectural skills and sense of historical drama he could capture the magnificence of a state occasion better than any of them. He also conceived a subtle plan …

> It struck him that to introduce portraits into this work would give it interest. Nothing daunted by his want of experience or reputation in this line of art, but feeling he could do that, or anything else if he tried, he set about writing to all the distinguished personages that attended this interesting ceremony, hoping to induce them to sit to him. And his hope was realised.
>
> He succeeded far beyond his most sanguine expectation. They came, they sat, they were painted. And when they came they saw the works hanging on the walls, or in his gallery, and admired them. Prince Albert, Earl Grey, Lord Howick, the Duke of Sutherland, one after the other sat, and Martin sold! [3]

Martin's *Coronation of Queen Victoria*, now at the Tate Britain, did not in fact sell for some years. When it did he received £2,000 for it. More importantly, he revived his reputation for stunning narrative canvases; many of his old works were bought and once more admired in the right circles. With returning energy, Martin's remarkable productivity also revived. In 1839 he exhibited six watercolours at the Royal Academy. He also began to rework *The Deluge* and it was while he was improving this 1826 painting that Prince Albert (like many a European prince, he was in town to court the most eligible lady in the world, his cousin) paid his first visit to Martin's studio.

King Leopold had introduced Albert to much that was modern and interesting in Europe. He was no ingénu. Albert had an eye, or at least the pretension of an eye, for Gothic art. Like his Uncle Leopold and many others Prince Albert found Martin sympathetic and engaging company. He suggested that *The Deluge* might form part of a flood trilogy, and in anticipation of his marriage to Victoria in 1840 he commissioned Martin to paint *The Eve of the Deluge* for Buckingham Palace. The *Assuagement of the Waters* was commissioned by the Duke of Sutherland, a potentially powerful and hugely wealthy new patron. John Martin's artistic rehabilitation, which had taken less than a year, was almost complete by the end of 1839. Now he returned to the Old Testament for inspiration and exhibited *The Destruction of Tyre* and *The Flight into Egypt* within another two years. In 1843, as if to mark the end of a process of healing, he exhibited *Solitude*, now on display at the Laing Art Gallery in Newcastle. It is an ethereal landscape of loneliness and calm reminding one almost of the pre-apocalyptic serenity of *Clytie* (which hangs next to it) but suffused with a sense of sadness and stoic acceptance.

This recovery from the depths of a sublime despond, bizarrely triggered by the deaths of his brother and nephew, brought John back into a more social world which had for some years been lost to the Martins. The household at Allsop Terrace was now substantially managed by the Martins' oldest daughter Isabella. John, when he was not painting, and Leopold, when he was not attending his duties at the Stationery Office, resumed their long rambles across a city that grew in size and bustle year by year.

At some time during the winter of 1838–9 they paid a visit to Turner at Queen Anne Street, no more than a ten-minute walk away. This was the domestic part of the complex which included Turner's studio and his gallery on Harley Street. At this time, his father having died, Turner was living a curious double life. The house in Queen Anne Street was run by a housekeeper, Hannah Danby, niece of Turner's former lover Sarah Danby by whom he had had two daughters. His father had been dead for nearly a decade; his mother had died in Bedlam more than thirty years previously.

For the last ten years Turner had also kept (or been kept by) another mistress, a wealthy widow of Margate named Sophia Booth. Sometime during the 1830s he installed her in a house on the River Thames at Chelsea where they lived together incognito as man and wife. Neighbours called him 'Admiral Booth' and had no idea that he was the great painter. Many of his friends knew nothing of this separate existence.

At the time of John's and Leopold's visit in 1838 Turner was painting one of his most famous canvases, *The Fighting Temeraire*. This grand old survivor of the battle of Trafalgar had just been towed up the River Thames by a steam tug to be broken up and Turner, returning from a visit to Margate, had been struck by the irony of the scene. He regarded it as one of his favourite works.

Leopold Martin paints an evocative picture of Turner at work in his sixties in a house that had been designed, but almost never used, for grandiose receptions ...

> The house was a gloomy, detached, five-windowed, large doored abode. An extensive studio occupied the back of the residence. We found the great painter at work upon his well-known picture 'The Fighting Temeraire'. Mr. Turner hardly struck one as a man who was producing works so full of poetry and art. His dress was certainly not that of a refined gentleman and painter. A loose body-coat, very open side-pockets, with a dirty paint-rag stuck in one of them; loose trousers, unbraced, and hanging under the heels of his slippers; a large rosewood palette on his thumb, with a very big bunch of brushes of various sizes in his hand; and a rather old hat on his head – such was JMW Turner at work. The studio was dark and gloomy, in every way like that of an untidy man ...[4]

Leopold's father's studio, in contrast, was light, airy and highly organized: a commercial, purpose-built printing and painting establishment full of carefully maintained equipment and neatly labelled drawers of materials.

As luck would have it, on the arrival of his visitors the unwashed and unshaven Cockney master was about to walk over to his secret establishment on the river. He invited the Martins to accompany him and so the party set off on foot, enjoying a 'very pleasant ramble' through Hyde Park, Brompton, and the market gardens of South Kensington to Turner's 'small place' at the far west end of what is now Cheyne Walk, where the Martins themselves eventually came to live.

Mr. Turner introduced us to a small six-roomed house on the banks of the Thames, at a squalid place past Lindsey Row[5] near Cremorne House. The house had but three windows in front, but possessed a magnificent prospect both up and down the river. With this exception, the abode was miserable in every respect. The only attendant seemed to be an old woman[6] who got us some Porter as an accompaniment to some bread and cheese. The rooms were very poorly furnished, all and everything looking as though it was the abode of a very poor man. Mr. Turner pointed out with seeming pride the splendid view from his single window, saying 'Here you see my study – sky and water. Are they not glorious? Here I have my lesson night and day.' The view was certainly very beautiful, but hardly of that description one would have expected the great Turner to glory in. Effect was all that was required. Mind gave the poetry of the picture.[7]

Turner's internalized world, unattractive as it seemed to London's self-conscious gentlefolk, allowed him to envision new realities quite beyond even John Martin's soaring imagination. His maturing sensitivity to texture, to atmosphere and colour is breathtakingly modern; his contemporaries, Martin included, it had begun to disturb. Leopold for one believed that 'he no longer sympathised with nature, but coquetted with her'.

Fortunately for Turner's reputation the young John Ruskin was soon to proclaim his genius in *Modern Painters*, first published in 1843. They first met in 1840 when Turner was 65, Ruskin 21 – himself a budding artist, though from a more genteel background than Turner. His father was a sherry merchant: refined taste ran in the family. This was the year that Turner, some thought, had gone too far. His *Slavers Throwing Overboard the Dead and the Dying* which commemorated the infamous and morally supercharged crime of Captain Luke Collingwood[8] in 1781 is not just impressionistic but phantasmagorical, with its saturated sunset and weird realization of oceanic monsters. Many now thought him mad. It was Ruskin's adamantine appreciation of his genius which founded his modern reputation. It was also, perhaps,

Ruskin's dismissal of John Martin that wrecked *his* reputation. In one essay he wrote that Martin's 'chief sublimity consists in lamp-black'.[9] In the manuscript for his *Stones of Venice* Ruskin castigated ... 'Workmen such as John Martin, whom I do not regard as painters at all. Martin's works are merely a common manufacture, as much makeable to order as a tea-tray, or a coal scuttle ...'[10] Martin may have been hurt by such criticism, but he was at least used to it. His friends the Hunts, Michael Faraday, the Brunels, Caroline Norton, had all suffered worse. In any case, there was surely something nobly Promethean about the prolonged torment of the chained martyr; and by succeeding in the face of both their critics and an establishment in denial of the march of progress they cast off their chains. Many of them joined the establishment, Martin included. The Titans might now live safely on the slopes of Mount Olympus: its groves as it were gaslit and paved by their own exertions.

In 1835 Frederick von Raumer had been taken aback by the value which Londoners placed on time. By 1840 it had become a national obsession. A steam passage could be taken in Isambard Brunel's SS *Great Western* from Bristol to New York. The rail journey from London to Birmingham took a mere five hours, and George Bradshaw had published his *Railway Companion*, the world's first rail timetable which included lists of all gradients and levels and two folded maps. Moritz Herman von Jacobi had built the first working electric motor based on Faraday's experiments and theories of electromagnetic induction. And Wheatstone and Cooke had patented the electric telegraph which ran messages alongside railway lines at something close to a third of the speed of light and would eventually connect continents via submarine cables.

Charles Wheatstone, now Professor of Experimental Philosophy at King's College, continued his researches into the transmission of light, sound and electricity throughout the 1830s. He experimented with electromagnetic motors. His fascination with measurement led to the invention for which all modern electrical engineers revere him: the Wheatstone Bridge, a device for measuring an unknown electrical resistance. He had been playing with the transmission of electrical signals more for the sake of scientific curiosity than because he saw economic potential in a telegraph. He gave up lecturing, for in truth he was a very bad lecturer, his 'wonderfully rapid utterance', as Leopold Martin had described it, a bar to public speaking. If he had to deliver scientific papers he got his friend Faraday, the most wonderful lecturer of his generation, to deliver them for him. Although Wheatstone had established his own reputation as a brilliant experimental and applied scientist, at a decade younger than Faraday he remained in awe of his friend's formidable intellectual powers. According to one former student, after a visit from Faraday, Wheatstone 'would sink into a reverie in which he seemed unaware of anything around him'.[11]

Wheatstone was not the only researcher to be thinking about a means of sending telegraphic messages using electricity. Samuel Morse was experimenting with a system in the United States, though his first practical apparatus was not installed for another decade and more. In 1816 Sir Francis Ronalds had built a prototype in his London garden – and the young Wheatstone had seen it.[12] Sommering in Munich and Moncke in Heidelberg were also working towards realistic designs in the 1820s and 1830s. The first practical machine is thought to be that of Baron Pavel Schilling in St Petersburg in 1832, though it was not immediately developed commercially.

Wheatstone's advantage lay both in his intimate knowledge of the latest research from across Europe and in his relationship with Faraday, who could be counted on to come at both theoretical and experimental problems with characteristic clarity and depth of knowledge. So it was to Wheatstone, very sensibly, that William Fothergill Cooke turned as a collaborator to build his electrical telegraph. Cooke was born in Ealing, London in 1806, the year that John Martin first arrived in the capital. His father was a medical doctor and professor of anatomy. At the age of 30, whilst conducting his own medical studies, William happened to see a demonstration of Schilling's telegraphic machine given by Moncke in Heidelberg.

Inspired by its possibilities Cooke took himself off to read Mary Somerville's 1835 *On the Connexion of the Physical Sciences*, gave up medicine immediately, and within three weeks had built a machine that used electromagnetic induction to deflect a needle: a galvanometer. This principle had been demonstrated by Oersted in 1820 in the experiment that had set such a dangerous hare running in the bowels of the Royal Institution and led to Faraday's temporary disgrace. Its use was a sine qua non of practical telegraphy.

Cooke, unhindered by a decade of competing research just as Davy had been unhindered through ignorance of the phlogiston debate, saw an opportunity to make himself very wealthy. Arriving in London at the beginning of 1837 he went to see Michael Faraday, who thought his ideas practical but without commercial potential. Persisting, Cooke spent four hundred pounds of his own money having improved examples of his machine made up and was then directed by the secretary of the Royal Society, Peter Mark Roget (of Thesaurus fame) towards Wheatstone. They met in February and agreed to become partners.

Wheatstone provided the science and practical knowledge of electricity; Cooke contributed the zeal of a convert and his newly discovered business drive. The essential technical difficulty was that resistance in Cooke's wires soaked up the weak currents he was able to generate before they could reach the electromagnetic deflector at the other end of the apparatus. Other obvious problems that needed attention were the requirement for messages to be sent in both directions, for an alarm to sound at the receiving end when a message was being transmitted, and technical restrictions on the number of symbols that could be represented using what was essentially an on–off switch.

Wheatstone had already worked on his own telegraphic devices; he had the depth of scientific knowledge to understand the nature of these problems and to suggest avenues for experimentation. So he suggested some promising lines for research and Cooke experimented. By July 1837 the two men had filed a patent for the first practical electric telegraph.

Although it underwent many modifications and improvements from very soon after it was built, the essentials of this first machine were straightforward – that is, they were straightforward to use; the technical issues which had to be overcome were immense. Five wires ran from one terminal to the other. Connected via Wheatstone's own 'permutating' circuit, they used electromagnetic impulses to deflect five needles. Pivoted at their centre like a compass, each needle could occupy three positions: left, right and centre (or, northwest, northeast and north). The centre position, in which the needle pointed vertically to nothing, was a 'neutral'. If the needle pointed to NW it indicated one letter; to the NE, another. Letters were arranged in pyramid formation on a diamond-shaped board in which the five needles lay in a line horizontally along the middle: there were four letters on the first row, three on the second, two on the third, and one at the apex of the pyramid – ten in all. Below the line of needles was another, inverted pyramid of letters, so that 20 letters in all could be indicated. The first letter on the left angle of the upper pyramid was 'H'. To indicate this letter the first needle was deflected to the NE to point towards it; the second needle pointed NW towards it. The next letter to the right was 'I'. To indicate this letter, the second needle was deflected to the NE, the third to the NW. The first and third needles pointing respectively NE and NW indicated a letter on the second row: 'E'. Letters on the lower pyramid were indicated by the needles pointing to the SW and SE. With five minutes' instruction anyone could send or receive a message on this system, allowing for the fact that the letters Z, X, J, C, Q and U could not be used. A bell could be sounded at either end to alert the operator to an incoming message or to acknowledge receipt of a message. It was beautifully simple.

By the beginning of August the intrepid Cooke had persuaded Robert Stephenson to test the new machine between Euston and Camden Town on his London to Birmingham railway. Stephenson initially saw only a limited application on his line, essentially as a signalling system. Isambard Brunel was more enthusiastic, however, allowing Cooke to install improved versions of the telegraph between Paddington and stations as far as west as Slough. With each new version the telegraph was improved, its design more efficient, its installation costs cheaper, its performance more reliable.

Samuel Morse arrived in London in 1838, the year following the Cooke and Wheatstone patent, to file his own telegraphic machine, only to find that he was too late. He took consolation in attending

Queen Victoria's coronation. He was not among those asked to sit by John Martin.

Wheatstone might have been temperamentally unsuited to a business partnership but he did enjoy collaborating with clever, like-minded people. His relationship with Cooke eventually fell foul of commercial jealousies. Others were more mutually sympathetic. One man whose ideas he readily supported was Rowland Hill, with whom in about 1833 Wheatstone formed a society for the promotion of scientific inventions. Hill was the supreme administrator of his day. His small fame as the 'Penny Black' man masks a series of achievements that stand alongside the telegraph, the railways and the steam printing press as Promethean epitomes. Like his fellow firebringers he was a passionate advocate of reform, education, humanity and science. Like so many of them his ideas matured over a long period and had to overcome bitter opposition from entrenched attitudes in the political establishment.

Rowland Hill's father was a teacher, a friend of Tom Paine and Joseph Priestley. Hill the younger was born in Kidderminster in 1795, like William Blake and Jonathan Martin a third son. He spent his early childhood confined and lying on his back because of a spinal condition. His father fed him literature and the principles of social reform. Having recovered from illness sufficiently to become a pupil at his father's school, Rowland also began to interest himself in science. He combined a love of mathematics and astronomy with skill as an artist and mechanical inventor.

As a teenager he was employed by his father, a follower of the Lancasterian method of teaching, to coach the younger pupils.[13] They established a species of self-government and a pseudo-legal system of trial and punishment in which pupils sat as jurors on their fellows. Discipline was severe. Corporal punishment was replaced by the imposition of menial tasks, a system which had already been proved effective by Lord Collingwood in the much harsher environment of the Royal Navy.

In about 1820 the Hills built a new school at Hazelwood, near Birmingham, to test Utilitarian methods and theories on a larger scale. Two years later they published an influential paper entitled 'Plans for the Government and Liberal Instruction of Boys in Large Numbers Drawn from Experience'. It was not just the theoretical and moral structure of the Hazelwood school that impressed educationists like Jeremy Bentham and Henry Brougham. It was also superbly equipped. There was a theatre in which pupils staged their own plays every term – Hill painted the sets and his sister made the costumes. The workshops were provided with small steam engines and with tools and equipment for experimenting with electricity. There was an astronomical observatory, an art studio, a music school and a modern printing press: all in all it sounds rather like an expanded version of Shelley's rooms at Oxford. The school began to attract interest from around the world. After a decade of success the Hills moved to larger premises at Bruce Castle School in Tottenham, North London. Charles Babbage sent his sons Herschel and Charles there on the advice of Jeremy Bentham and never regretted it (whether or not his sons did is another matter).

Rowland Hill went to see Babbage at the time when he was beginning to work on his Difference Engine.[14] Babbage envisaged many applications of the calculator, including its use in verifying prices. In his *Economy of Machinery and Manufactures* (written in 1832; Karl Marx read it in 1845 and it changed his life) he had argued for a uniform postal rate to be calculated by his engine. During their discussions Hill was a ready convert to his ideas. It was Babbage who enthused Hill with his postal schemes, and Babbage who introduced him to his method of operational research. Hill had spent all his adult life as a teacher. Now he began to see that his talents might go further.

In 1833 Rowland Hill, now nearly forty, abandoned Hazelwood. He began to collaborate with Charles Wheatstone (they may have been introduced by Babbage) on the promotion of inventions; he devised a means of sorting letters in mail coaches; he designed clocks which told the time by reference to the stars, and a rotary printing press. He had embarked on a journey only too familiar to men like John Martin, Samuel Taylor Coleridge and Humphry Davy, restless explorers and visionaries endowed with almost too much talent, touching on everything but rarely settling.

In 1835 Hill, now financially embarrassed, took a post as Secretary to the South Australian Commission. His remit was to organize non-convict emigration to the area around Adelaide and he set about reforming an essentially unregulated means of disposing of Britain's surplus population. One aspect of Britain's relationship with her new colony he found particularly grotesque: the cost of sending a parcel to South Australia could be as high as £20–25. It was an example of a postal system which was notorious for iniquity, corruption and labyrinthine complexity. Remembering his discussions with Babbage, Hill now envisioned a reformed postal system based on Utilitarian principles: the greatest good for the greatest number.

The Royal Mail primarily served Britain's wealthy, and in particular her parliamentary members whose generous postal allowance they abused shamelessly. Free parcels, franked under their name, were known to include items like pianos, cows and maidservants.[15] Public letters were subject to an initial charge depending on the number of sheets they contained (postmasters 'candled' them, holding them up to a light to count the number of leaves of paper). An additional charge was added depending on the distance to the recipient, who had to pay to receive the letter. Parcels attracted a further charge depending on their weight. Because many recipients of mail could not afford to pay the postal fee, a mountain of dead letters accumulated each year – worth an estimated one hundred and twenty thousand pounds in lost revenue to the Exchequer.[16]

In the year that Hill took up his post with the South Australian Commission Robert Wallace, Member of Parliament for Greenock, persuaded the government to set up a committee to look at the postal system. Not for the first time, mercantile industrial interests clashed with those of the landed aristocracy. Merchants saw a reformed postal service as highly desirable, especially if it was combined with the speed of the spreading rail network. Establishment Tories saw only a further diminution of their privileges and the ever-present spectre of Radicals using speedy communications to foment Jacobinism. Had not Henry Brougham used the Mail in 1832 to force the government's hand during the Reform Bill crisis? Were not bankers' clerks absconding from their employers at the rate of 20 miles an hour in the Stephensons' locomotives?

In 1837 Hill, having spent two years analysing the government Blue Books which detailed departmental finances, and having compared the British postal service unfavourably with that of France, published a hundred-page book called *Post Office Reform: Its Importance and Practicability.* He sent the first copy to the Prime Minister, Lord Melbourne. Hill's theory was that by introducing a uniform low rate of postage across Britain, increasing only with the weight of a letter, firstly administration costs and corruption would be reduced; and secondly a vast increase in the volume of mail would net the government hundreds of thousands of pounds in extra revenue. He argued that there should be daily deliveries to every town and village; that houses should have letter boxes cut in their doors, and that there should be posting boxes in convenient places for the public to send letters. A single folded sheet should be used to write on and the postage should be paid by affixing a small label, or stamp, to the exterior of the letter next to the address. In all its essentials, Hill had in one sweeping canvas portrayed the modern mail service. He had also fired a volley at the establishment, and they did not like it.

Hill was interviewed by the Chancellor of the Exchequer, the exotically named Spring Rice. He was patiently heard by the Commission. And then ... nothing. The Post Office was run more or less as a party sinecure. Change was unthinkable. The opposition of the Postmaster General was implacable: 'Of all the wild and visionary schemes I ever heard of, this is the most extraordinary', he wrote.[17]

Hill responded by circulating his report to a more sympathetic audience: the banks, business houses and newspapers in whose economic interests a reformed system would work. *The Times* and the *Spectator* wrote leaders in support of the scheme. Two thousand separate petitions were sent to the government demanding that Hill's plan be introduced. Reluctantly (a man like Melbourne could never enthuse over something as trivial as sending a letter) the Prime Minister appointed a new Select Committee to look at Hill's proposals. The Committee recommended full implementation of his plan by the single casting vote of the Chairman. Despite bitter opposition from Lord Lichfield, the Postmaster-General, Melbourne bowed to pressure from a deputation of 150 Members and in July 1839 the Chancellor announced the introduction of the Penny Post. In August, the Postal Bill was passed.

Melbourne took a much more personal interest in another bill passed that year: the Custody of Infants Act. In the aftermath of his case against the Prime Minister, George Norton had taken the three boys to Scotland where they were kept by his relatives out of the way of their mother. He entered into a sleazy correspondence with his estranged wife, alternately accusing and pleading with her. He impounded Caroline's books, jewels and clothes against her debts, real or imagined. He told her she might see the boys again if she would come to a mutually satisfactory financial arrangement – in other words, pay him off with the earnings from her poetry and editorial employment. Desperation to be reunited with her sons fought a tormenting battle with pride, anger and a sense of bitterness. Norton's lawyers teased her; the Sheridan family begged her not to agree to anything without absolute assurances that she would be able to see the children regularly. On occasions she gave in, only for new conditions to be attached to her access to the boys. She was furious at being duped but adamant that she must pursue any avenue to be reunited with them.

Caroline's relationship with Melbourne had suffered during the trial. They could not be together; their intimacy was at an end. At first he thought she was angry with him; a tentative correspondence was resumed after she assured him that her silence was caused by her misery, that she did not blame Melbourne at all. Gradually, she began to ask his advice once more, and he gave it readily. But he knew she was being humiliated by Norton's vacillation. Melbourne did not trust Norton and he told her so in a letter full of compassionate rebuke and self-pity which began by trying to persuade Caroline not to write imploring letters to her husband.

> Every communication elates him and encourages him to persevere in his brutality. You ought to know him better than I do, and must do so. But you seem to me to be hardly aware what a gnome he is,

how perfectly earthly and bestial. He is possessed of a devil,
and that, the meanest and basest fiend that disgraces the infernal
regions.[18]

Melbourne knew very well that Norton's overriding interest was money
and that he was being egged on by his family even at the expense of
his children's happiness. He also knew his correspondent. He realized
that she must be struggling financially but he knew only too well the
intensity of her emotions and fiery pride. In July 1836 he broached
the delicate subject and exposed his own emotions so that Caroline
might be absolutely assured of his loyalty to her. If this is not a love
letter, then such a thing does not exist.

> I have never mentioned money to you, and I hardly like to do it
> now; your feelings have been so galled that they have naturally
> become very sore and sensitive, and I knew how you might take it.
> I have had at times a great mind to send you some, but I feared to
> do so. As I trust we are now upon terms of confidential and
> affectionate friendship, I venture to say that you have nothing to
> do but express a wish, and it shall instantly be complied with. I miss
> you. I miss your society and conversation every day at the hours at
> which I was accustomed to enjoy them; and when you say that your
> place can easily be supplied, you indulge in a little vanity and
> self-conceit. You know well enough that there is nobody who can
> fill your place.[19]

This letter was written less than a year before William IV's death
brought Melbourne into contact with the young queen. After Victoria's
accession Melbourne's political and emotional energies were increasingly directed towards her and once again Caroline began to believe that
he was neglecting her. Fortunately, he was not Caroline's only friend,
however often she must have felt excluded from the circles which she
formerly enjoyed. Apart from the impressive ranks of Sheridans (her
half-uncle Charles took her to live with him in Green Street at the
northeast corner of Hyde Park) there was Sam Rogers, who remained
devoted to her cause. There were still admirers and one especially,

Sidney Herbert, whom many believed became her lover for a time. And she found a most welcome ally in Harriet Sutherland-Leveson-Gower, Duchess of Sutherland and wife of John Martin's powerful new patron.

The duchess was young and beautiful, peerlessly well-connected. She later became Mistress of the Robes to her friend the queen. After the Norton–Melbourne trial she invited Caroline to ride in her carriage around Hyde Park, as ostentatious a display of support and approbation as could be conceived at such a fashion- and status-conscious period. This was an act of considerable generosity, a passport to a more close-knit and discreet circle that Caroline was able to enjoy and profit from in the future.

Another close friend at this time was Abraham Hayward, translator of Goethe's *Faust* and noted gastronome. Caroline approached him because he had parliamentary connections which could not be construed as having anything to do with Melbourne. She wanted him to help her frame legislation that would assure contact between mothers and their children. She wrote a pamphlet, *Separation of Mother and Child by the Custody of Infants considered* (1837) which argued the appalling iniquity of the current law and its potentially devastating effects on children and mothers alike. Since a married woman had no independent legal existence, she had no right to the custody of her own children, any more than she had a right to her own property: she could not legally have any property – it belonged to her husband.

In the early part of 1837 George Norton brought the three boys Brinsley, Fletcher and William down to London. Caroline saw them briefly. George pleaded with her to return to him and she came close to giving in before realizing that he was, as before, stringing her along in order to win himself a favourable settlement. There were more furious arguments, scenes of terrible distress as Caroline tried to see her boys in secret and had them literally torn from her arms by George's relatives. That year her bill almost came before Parliament but fell foul of a postponement. In May 1838 the bill was reintroduced and passed the House of Commons before it was thrown out by the Lords. Henry Brougham, whom she had unwittingly antagonized during the early part of Melbourne's ministry, was its most vociferous opponent.

In the summer of 1839, after much lobbying and an unsavoury battle of words in the Whig and Tory press, Caroline's Infant Custody Bill was passed by both Houses. Melbourne had once again presided unenthusiastically over a landmark social reform: the first women's rights legislation in British political history. Nevertheless, he did not think it would do Caroline much personal good, and nor did it.

Caroline's stock was slowly rising once more. She had her bill. She had at last been received at court. Her literary career was at its height and she now revived her career as a hostess. She corresponded with old friends like Mary Shelley and arranged introductions between foreign diplomats and men like Charles Dickens, then enjoying huge success with *Oliver Twist* and *Nicholas Nickleby*. She had regained her position in society, but still she did not have her children. George Norton's response to his estranged wife's enhanced moral and legal position was once more to remove his children to Scotland, where the Act had no force.

In January 1840 the first Penny Post came into operation. There were queues at post offices across the country. Rowland Hill's old drawing-master stood all night outside the office at Birmingham in order to be the first to send a penny letter to his former pupil by way of congratulation. London's post office handled a hundred and twelve thousand letters on the first day. Hill had been given a junior temporary appointment in the Treasury by Melbourne, reluctant to ruffle too many of the Postmaster-General's feathers. Now he gave up his work at the South Australia Commission and set about implementing the organizational and technical innovations that his visionary system entailed. In May 1840 he introduced the Penny Black and Twopenny Blue, the first adhesive postage stamps, which bore the portrait of the queen.[20]

Having now found her feet and reached her 21st year, the queen began to assert sovereign independence from a loving but overly attentive Uncle Leopold and a domineering mother. Melbourne had given Victoria the confidence to become her own queen and she now intended to exercise her authority. She had no shortage of concerns. The new postal system had very rapidly since its introduction been exploited by a coalition of popular movements dominated by Chartists and the Anti-Corn Law League. These now threatened to reopen wounds unhealed by the Reform Bill.

The People's Charter was the reaction of a working class that had been cheated by the Reform Bill and knew it. Robert Owen's trades unions had failed to improve the lot of the labouring poor who felt the brunt of economic recession in the late 1830s. Parliamentary reform had been a defensive measure to appease the manufacturing middle class at the expense of their workers who now, in righteous anger, demanded what they had wanted all along: universal adult male suffrage, annual parliaments and the secret ballot. The Chartists sought a wage for Members of Parliament so that ordinary men might represent their constituencies. They also demanded that constituencies represent equal numbers of electors. It all seems so reasonable. But to those, the newly wealthy and self-satisfied middle classes, for whom the vision of equality was a nightmare, Charles Dickens' lurid portrayal of industrial poverty and political agitation was only too believable:

> bands of unemployed labourers paraded the roads, or clustered by torch-light round their leaders, who told them, in stern language, of their wrongs, and urged them on to frightful cries and threats; when maddened men, armed with sword and firebrand, spurning the tears and prayers of women who would restrain them, rushed forth on errands of terror and destruction, to work no ruin half so surely as their own.[21]

The People's Charter was presented to Parliament in 1839 by William Attwood, former leader of the Birmingham Union and now himself a Member in the House of Commons. It carried more than a million signatures. The Chartist newspaper, the *Northern Star* (founded in 1837) was distributed by the new Penny Post.

Melbourne did not believe in any of the measures demanded by the Chartists. He was in any case preoccupied. He was forced to resign as Prime Minister in May 1839 because he had very nearly lost a vote on colonial issues (trouble in Canada and Jamaica). He advised the queen to send for the Duke of Wellington, knowing that the duke would recommend Robert Peel to Her Majesty. Peel's social awkward-ness and his request that some of the Queen's Ladies (all of them married to Whigs) should be replaced by his own appointees drove the queen into a fury. She refused and appealed to Melbourne to take up office again. For reasons of personal loyalty rather than love of power he agreed, although by now his health was in decline and he longed to retire.

The queen's display of petulance went down well with the public. Melbourne's Whigs earned themselves a breathing space and they were able to portray the Chartist petition, as they had done its predeces-sors, as undesirable, revolutionary and un-British. The Chartists licked their wounds and waited for another chance to attack. Melbourne soldiered on, increasingly uninterested in his cabinet, his opponents and the demands of the public for reform. Now that the queen had decisively found her feet he spent much of his energy trying to soften her attitude towards Peel and his Tories: a partisan on the throne was a dangerous thing, as Melbourne knew from his own deep knowledge of English constitutional history. But above all he was tired. He had never possessed the restless visionary energy of the Prometheans. The end, when it came, was a relief.

Melbourne's Government succumbed in 1841 to its own quiet fatigue and the clamours of the Anti-Corn Law League. The Corn Laws had been an object of hatred since their introduction during the war against France. They underwrote landowners' interests against those of workers who could not afford to buy bread. In an increas-ingly industrialized economy this farming subsidy infuriated manu-facturers who had to compete in an open market, taking its booms and busts as they came. In periods of economic recession, like that in the years leading up to 1840, the laws provided a focus for discontent whose causes and remedies were otherwise perhaps too complex to see clearly.

The Anti-Corn Law League pounced on Rowland Hill's postal reforms. A rapidly expanding rail network allowed them to commu-nicate quickly and cheaply: above all to organize efficiently. Their news-papers, the *League*, the *Economist* and the *Sun* were supported by a middle class with far greater resources than the Chartists. They argued that the end of farming subsidies would not just open British markets to foreign producers but also expand British manufacturing and export opportunities to the wider world where cheap mass-produced goods were very much in demand.

Melbourne was not a natural Whig. He had never been a radical. His politics were instinctively conservative. He believed in the patronage system and he believed that, despite its flaws, the rule of landowners was right, just and effective. It had always worked, and the less it was tampered with the better. Increasingly, though, both opponents and colleagues were won over by the idea of repealing the Corn Laws – not from a compassionate wish to feed the poor, but for pure economic self-interest. Whatever else it had done, reform had increased the number of manufacturers who either themselves sat in Parliament or who controlled urban boroughs. Above all, Melbourne wished to prevent a crisis like that of the early 1830s. His temptation was to offer a free vote, but his political head told him that he must finally intervene one way or the other.

In May 1841 the government was defeated over a compromise budget which proposed a fixed duty on corn imports. Peel proposed a vote of no confidence which was carried by a single vote: 311–310. Parliament was dissolved and in the General Election that August the Tories, under Robert Peel, were returned with a majority of 90. Peel, the son of a textile baron, was a natural champion of manufacturing interests but his party still wore protectionist colours. Even so, the Free Trade writing was on the wall and the widely despised Corn Laws were finally repealed between 1846 and 1849.

Melbourne, in some respects the last eighteenth-century Prime Minister, took his leave of the queen refusing, as he always had, the honours which she offered him. Now released from his yoke he was to be seen strolling along The Mall to the House of Lords at whatever time it pleased him to attend, smiling to himself. Discreetly, very discreetly, he could occasionally be seen having dinner with Caroline Norton.

CHAPTER FIFTEEN

Judgement

HOW HAVE THE PROMETHEANS been judged by history? As individuals they have enjoyed mixed post-mortem careers; as an identifiable entity they remain unrecognized. It is true that their faces appear on bank notes and postage stamps, their works are 'classics', their monuments revered as memorials to a certain sort of indefinable 'greatness'. But historians have generally neglected them as a coherent force for change despite the tangible social, cultural and intellectual ties that bound them together in the cause of self-enlightenment, emancipation, reform and profit. They did not ride on horseback to meet once a month by the light of the full moon as an earlier genera-tion of visionaries had; they did not need to, for between them they invented electricity, gaslight, telegraphy and the railways. They had no name for themselves, as the members of the Lunar Society, the Clique or the Pre-Raphaelite Brotherhood did. They were not, self-consciously, a group of any sort.

The British establishment, Tory and Whig alike, were wrong-footed by the Prometheans. They were not landowners. They wielded no power but that of the printing press, the laboratory, the workshop and the canvas. They did not throw stones at kings or burn hayricks. Their militancy was, for the most part, intensely personal. They could not, therefore, be stopped. They were the future, seeping under the door like an incoming tide.

Judged by their own aspirations they might be said to have half-succeeded; to have designed the tools with which to build an earthly paradise but to have left off construction at foundation level. Along with their counterparts in Europe and America Britain's Prometheans had elucidated the principles of electricity, modern chemistry, mass-manufacturing; of cheap and rapid transport and communication; of the computer; of social welfare and equity. They had experimented with the principles of republicanism, watching its practice with intense interest and occasional horror or admiration. They had won for themselves a social and physical mobility unthinkable to their parents. They had witnessed the abolition of slavery. They had broken free from the chains of romanticism and pointed the way towards positivism and the rational. Their works were enjoyed by a majority of the populace thanks to advances in printing technology, production and distribution. They had written the first modern novels and taken literary criticism to unparalleled heights. Their best poetry has hardly been bettered. Above all, they had conferred upon the children of a new generation a belief in self-liberation regardless of rank. It was a belief, suppressed in later Victorian Britain, that found its greatest expression in the so-called American dream (though even the Americans were required to fight a civil war to establish its legitimacy). It is no coincidence that New York's Rockefeller Center boasts a gilded bronze *Prometheus* fountain by Paul Manship; that the Statue of Liberty stands in New York's harbour as a beckoning icon for refugees fleeing poverty and repression in the Old World.

The artistic, scientific, engineering and literary legacy bequeathed by the Prometheans is entirely recognizable as the basis of a modern society. And yet, they had not achieved Shelley's dream of an equilibrium between institution and opinion, nor yet William Godwin's perfectible self-governing society. What generation could? The reformed Houses of Parliament, since 1834 not even in existence as a physical entity and still not rebuilt,[1] were closed to ordinary men and women and subject, as they always had been, to the vested interests of powerful families who had wielded power time out of mind. The health of Britain's expanding population was deplorable for want of sanitation, clean water, medical care, factory legislation and proper housing. London, supposedly the world's greatest city, still poured the filth of a million homes into the atmosphere and into the tidal waters of the Thames.

The generation which followed the Prometheans was just as angry as they had been; more so perhaps, because their fathers and mothers, elder brothers and sisters had fomented their aspirations. They began to believe that the freedom of thought, movement and ambition which had been won by their parents were rights to be defended against forces of repression which they saw threatening them on all sides. The ground that had been so painfully gained was held in a fragile but determined grip.

In 1841, as Sir Robert Peel, the first Prime Minister to be born of a manufacturing father, was returned to power in Britain (and as John Martin and Charles Wheatstone rode at 90 miles an hour on Brunel's Great Western Railway) a radical German philosopher and budding revolutionary was writing in his doctoral thesis for the University of Jena that Prometheus was 'the foremost saint and martyr in the philosopher's calendar'.[2]

That Karl Marx, the very image of an Old Testament prophet, explicitly recognized Prometheus as the god of the godless, the rejecter of servility; that he claimed man's 'self-consciousness as the highest divinity'[3] allows us to place him in a direct line which runs from Rousseau through Franklin and Jefferson, Bonaparte, Godwin and Shelley. He may even be said to have been the last true Promethean. Before Marx there was a natural order to things in which man played his part for good or ill in a drama of moral certainties. After him there was class struggle, international ideological war and a cynical belief that history must always repeat itself: the first time as tragedy, the second as farce.[4] The Greek Titan of insolent liberation was thus the inspiration for both the modern model of capitalism and its apparent antithesis, global communism. William Godwin, at least, would have appreciated the irony. What is shocking is that Marx, inheritor of an essentially eighteenth-century enlightenment mindset should, by taking Prometheanism to its *reductio ad absurdum*, not only build a complete theoretical language for Godwin's perfectible society but also draw up a blueprint for the monster created by the old anarchist's own daughter.

Karl Marx was born in Trier in the Rhineland in 1818. His father Heinrich was a lawyer of comfortable means, a literate man familiar with the ideas and works of the European Enlightenment: of Voltaire, Rousseau and Goethe. The son studied law in Bonn where he fell under the spell of Romanticism and alcohol, wrote poetry and received a slight wound during a duel. There was something of the Percy Shelley about him. In contrast to the lyric poet's parental hand, Marx's father persuaded Karl to transfer to the far more serious Friedrich-Wilhelms-Universität in Berlin. Here he was radicalized by the work of G. W. F. Hegel, whose recent death had deprived Germany of its foremost idealist philosopher.

In Berlin, Marx embraced atheism and the Promethean spirit and, turning to journalism, began to test the limits of political censorship in Germany and then France. His first articles were diatribes against censorship. An early piece for the reformist Cologne newspaper *Rheinische Zeitung* in 1842 was followed by a cartoon, in the style of Albrecht Dürer, of Marx chained to an old-fashioned printing press (surely a nod to Erasmus) with an Imperial eagle tearing out his liver. A more self-conscious identification with Prometheus cannot be imagined.[5] By the end of 1842 Marx had become the newspaper's anonymous editor (signing himself 'Prometheus') and been visited in his office by Friedrich Engels. Engels, like Jenny von Westphalen, whom Marx married in 1843, was a devotee of Shelley. Was Marx beginning to believe himself the embodiment of Shelley's call to arms in *Prometheus Unbound*, who would discharge the collective mind's lightning and restore the social equilibrium?

In Paris in 1844 Marx again met Engels, who had just published *The Condition of the Working Class in England in 1844*, and began to apply his skills as a philosopher to political economy. A year later Paris became too hot a revolutionary cauldron and they were ordered to leave. In 1845 they found themselves in Brussels, given sanctuary by John Martin's old friend and patron King Leopold. By the end of the decade they were in London, attracted to the agenda of a new radical group called the Communist League.

Britain had somehow reached the 1840s without being plunged into revolution. She survived that decade of turmoil also. Burke's old analysis of the British as cold, sluggish and reluctant to innovate was in some respects as true now as it had been 50 years before. But Britain's landowners, however conservative their values, were also pragmatic mercantilists, well aware that national wealth was a great antidote to revolution. They had learned to adapt. They had withdrawn their capital from slavery and sugar and were now, with great enthusiasm, sinking it into railways. If they had to give a little over corn prices they would, as they had done over parliamentary reform. Even the most radical Chartists were not proposing to relieve them of their country estates, their titles or their fleets of ships. At least, not yet.

It was Peel's failure to act swiftly enough on corn prices which brought his government down after five years. He was in truth lucky to survive his premiership at all, because in 1843 he was the target of an assassination attempt. Daniel McNaughton was a Glaswegian wood-turner in his late twenties. He had periodic delusions of persecution, including a sort of general paranoia concerning Tories. At some time in 1842 or a little before, he arrived in London and began to loiter in the streets around Whitehall. Some soldiers and policemen noticed his odd behaviour but did nothing to inquire into his purpose. In January 1843 McNaughton began to watch very closely the movements of the Prime Minister, whose private residence was in Whitehall Gardens. On the 20th he saw a man whom he believed to be the Prime Minister walking towards the Gardens. Taking a pistol from his coat he fired it at the man, who staggered to a nearby banking house. In the act of bringing a second pistol to bear the assassin was wrestled to the floor by a police-men and a passer-by. The second pistol went off but the bullet hit no one. The man he had shot, fatally as it transpired, was not Sir Robert Peel but his Private Secretary, Edward Drummond. The banking house he had staggered to was, in fact, his own.[6]

Daniel McNaughton was acquitted at his trial because although (like Jonathan Martin) he probably knew that what he had done was wrong, he had acted under a delusional impulse which was not consistent with his normal behaviour. Like Jonathan Martin, McNaughton was sent to be detained at Bethlem Hospital at Her Majesty's pleasure. Here he stayed until his removal to the new Broadmoor Asylum in 1864. The case caused an outcry in the press. George III's would-be assassin James Hadfield had, after all, been demonstrably injured in the head while serving as a soldier in the King's Service: had had an excuse; and he had not even succeeded in his aim of killing the king. Jonathan Martin had not killed anyone either. John Bellingham, who had succeeded in 1812 where Hadfield failed in killing his intended victim, Spencer Perceval, had been hanged as a murderer. Were all murderers now to be allowed to plead insanity after committing the ultimate horror? The queen (who had already survived attempts on her own life in 1840 and 1842) wrote to Peel, the intended victim, asking for a clarification of the law on insanity.

Henry Brougham, the successful defending counsel at Jonathan Martin's trial, rose in the Lords in March 1843 to say that he would be pleased to address the question of legislation if Their Lordships pleased. Brougham may have had his political ambitions thwarted, but the limelight still beckoned. Thus pressurized, the government appointed a panel of senior judges to offer an opinion which was to become known as the McNaughton Rules on criminal responsibility, based on a diagnostic test of five questions. They were to form the principal guidance on such cases to judges across the world for a hundred years and are regarded as a legal landmark in the treatment of mental responsibility.

Edward Drummond inadvertently lost his life in the cause of legal progress on mental illness at about the same time that Caroline Norton was paying an almost equally heavy price for her campaign to formulate rights for women. In 1841 she had been told that her three boys were to be sent from Scotland to an English school, one famous for its harsh physical regime. Hoping finally to be able to see them, she drove down to the school and was turned away. But now she had the law on her side. Again it was her friend Abraham Hayward who helped draft her application to the courts. Threatened with another expensive legal action George Norton gave way, allowing the boys to spend part of Christmas 1841 with her. Caroline must finally have believed that she was going to see them regularly, but after this brief reunion the boys were sent back to school and she was forbidden to visit them there.

The following September, while the boys were with their father at the Norton family estate at Kettlethorpe in Yorkshire, William, the youngest of the three boys, was hurt in a fall from a horse. A superficial scratch which he sustained on his arm became infected, probably with tetanus. Quite how the Norton household in Yorkshire was organized it is hard to say, but the severity of the infection was evidently not recognized; the boys seem to have been largely left by their father to fend for themselves in a cottage some way from the main house. When Caroline was finally sent for it was too late. She was met at the station and, asking if her boy was better, was told that he was dead.

> Dearest Mr. Rogers …
> I still feel stunned by this sudden blow. The accident happened here, and I have been sheltered here ever since, and do not leave till Thursday, when my fair young thing will be laid in the grave. The room here where he died … the room where there was so much hurry and agony, and then such dismal silence and darkness – is empty and open again, and the little decorated coffin is lying at his father's house (about two miles off) alone …
>
> It may be sinful to think bitterly at such a time; and at least I have not uttered the thoughts of my heart; I have choked them back, to spare pain to one who never spared it to me! But it is not in the strength of human nature not to think, 'This might not have happened had I watched over them!' [7]

Now, finally, Caroline won an agreement to have the two remaining boys stay with her for half the year. It was not the end of her war with George Norton – that continued on and off until his death in 1875. But she might now begin to build something of a relationship with her estranged sons Brinsley and Fletcher. Other wronged women might now fight for their rights with her legal and moral example before them.

Caroline returned to writing, publishing a series of successful novels. The 1840s were one of the most fertile decades for English writing and her books do not stand up to modern scrutiny in the face of what now seems like almost unfair competition. At the beginning of the decade her friend William Harrison Ainsworth had achieved huge success with his *Tower of London* and *Old St Paul's*, the latter with its unfortunate cameo appearance by Jonathan Martin as Solomon Eagle, the ranting pyromaniac prophet (the portrayal does not seem to have affected Ainsworth's friendship with John Martin in the least).

In 1842 *Punch* was first published, its most famous illustrator, John Tenniel, a graduate of the Martin household and brother-in-law of his friend Leopold Martin. Another mutual friend, Charles Dickens, produced *A Christmas Carol* the following year: a more readable though hardly more important work than John Stuart Mill's *System of Logic*, also published in 1843 (the year when the first part of John Ruskin's *Modern Painters* was completed). Engels published his *Condition of the Working Classes in England* a year later. In 1845 a future Prime Minister, Benjamin Disraeli, whom Melbourne had so patronizingly dismissed at one of Caroline Norton's parties, published *Sybil* (an unsubtle anti-Chartist melodrama) and Edward Lear's *Book of Nonsense* was published the year after that.

The year 1847 was a miraculous one for literature: *Agnes Grey*, *Wuthering Heights*, *Jane Eyre*, all brought out in an explosion of dramatic fiction from the Yorkshire family who counted John Martin as one of the inspirational figures of their childhood (their toy figure Edward de Lisle of Verdanopolis was an incarnation of the painter). Thackeray's *Vanity Fair* was also published in serial form, initially to cool reviews. Its cynicism required a new sophistication from an audience pampered by cardboard cut-out characters and soap-opera plots.

Uncharitable readers might have seen in Thackeray's Becky Sharp a fictional Caroline Norton; she would perhaps have been more fairly served by comparison with Helen Huntingdon, the wronged heroine of Anne Brontë's *Tenant of Wildfell Hall* which appeared in 1848 and which must stand as the great pioneering novel of feminism. It was not, however, the most important work to appear on the streets of London that year. Even John Stuart Mill's hugely influential *Principles of Political Economy* must take second place to the launch of the *Communist Party Manifesto*. It was released (initially in German) against a backdrop of impending continental revolution which must have seemed like the fulfilment of John Martin's apocalyptic predictions in *The Fall of Nineveh* and *The Fall of Babylon*:

> The Communists disdain to conceal their views and aims. They openly declare that their ends can be attained only by the forcible overthrow of all existing social conditions. Let the ruling classes tremble at a Communistic revolution. The Proletarians have nothing to lose but their chains. They have a world to win.
> WORKING MEN OF ALL COUNTRIES, UNITE! [8]

Written more than 70 years after Jean-Paul Marat's inflammatory *Chains of Slavery* was furtively distributed around the streets of Newcastle, it is not the message but the breathtaking modernity of the language which is so striking about the *Manifesto*, reinforced by the formality of Marx's academic philosophizing and Engels' skills as an economic analyst. The style is imperative, unrelenting, mechanical, the embodiment of what Thomas Carlyle had feared when he said that 'men are grown mechanical in head and in heart, as well as in hand'.[9]

> The modern labourer … instead of rising with the progress of industry, sinks deeper and deeper below the conditions of existence of his own class. He becomes a pauper, and pauperism develops more rapidly than population and wealth. And here it becomes evident, that the bourgeoisie is unfit any longer to be the ruling class in society, and to impose its conditions of existence upon society as an overriding law. It is unfit to rule because it is incompetent to assure an existence to its slave within his slavery, because it cannot help letting him sink into such a state, that it has to feed him, instead of being fed by him. Society can no longer live under this bourgeoisie, in other words, its existence is no longer compatible with society.[10]

The *Manifesto* was written in London because that was where the Second Congress of the Communist League (meeting in secret) was held towards the end of 1847. Marx and Engels were both present. Engels was already perfectly au fait with the historical context of the workers' struggle in England; he is probably to be credited with coining the term 'Industrial Revolution'. He knew Robert Owen's works and was familiar with the Chartists' newspapers. Marx had read Babbage's work on machinery and manufactures[11] and was well acquainted with the progress of political philosophy in Britain. That is not to say that either of them believed revolution was possible in Britain, even though the Interregnum of the mid-seventeenth century provided a gratifying, if short-lived, historical precedent. Nevertheless, it is entirely fitting that the *Manifesto* was written in the London of Godwin and Wollstonecraft, Shelley and Martin.

One small but successful revolt did take place in England in 1848. At the great Chartist meeting held in London in April John Everett Millais (a well-bred artistic prodigy) and his friend William Holman Hunt (son of a warehouse manager) declared that they were going to stage an artistic coup against the Royal Academy. Inspired by a coalition of thinkers which included Thomas Carlyle, John Ruskin and Robert Owen, and motivated to follow where anti-Academicians like Martin and Haydon had led, they sought a return to a pure form of painting in which truth to nature would triumph over frivolity and self-parody.

In the autumn of that year they met and befriended Dante Gabriel Rossetti, son of an Italian exile, and formed the secret Pre-Raphaelite Brotherhood. These original brothers, united more in name and spirit than in technique or subject matter, revived a sense of natural simplicity and vibrant colour which they saw in fourteenth-century Italian art. Much of their work seems naïvely romantic, if exquisitely beautiful. But the movement which they spawned, lasting until the end of the nineteenth century, also bound into artistic consciousness the themes of liberation, social justice and education which can be seen in the creative output of Ford Maddox Brown, William Morris and John Ruskin. They may not have seen themselves as Prometheans, but they were the inheritors of Prometheanism.

Political revolution did in fact come within weeks of the publication of the *Communist Manifesto* – not in London, but in Paris. King Louis-Philippe fled to London. In February, and again in June, the streets of Paris ran with blood as they had done 60 years before. A Second Republic was founded, though it failed within three years. Marx and Engels, sufficiently sanguine to look at the wider historical picture, were undeterred:

> Now and then the workers are victorious, but only for a time.
> The real fruit of their battles lies, not in the immediate result, but
> in the ever-expanding union of the workers. This union is helped
> on by the improved means of communications that are created by
> modern industry and that place the workers of different nationali-
> ties in contact with one another. It was just this contact that was
> needed to centralize the numerous local struggles, all of the same
> character, into one national struggle between classes. But every
> class struggle is a political struggle. And that union, to attain which
> the burghers of the Middle Ages, with their miserable highways,
> required centuries, the modern proletarians, thanks to railways,
> achieve in a few years.[12]

The Chartists' attempt to galvanize London's proletariat in imitation of their European brothers in 1848 was a miserable failure. The repeal of the Corn Laws two years earlier had partially drawn their sting (as it was meant to) and, besides, the Duke of Wellington's expert deployment of the police and army on the streets and bridges of the capital must have reminded every zealot armed with a stick or poker that this was the general who had never lost a battle. The ever-expanding railway network (5,000 miles of it by 1848) which allowed the proletariat to travel freely across the country in the company of their middle-class compatriots, was also creating the economic conditions for Victorian prosperity which emasculated Britain's revolutionaries indefinitely.

The technological and engineering persistence of the Prometheans was exploited by a new wave of bourgeois entrepreneurs of varying shades of probity. The age of the engineer-cum-railway magnate was already passing. By the time of the upheavals of 1848 one man effectively controlled something like a third of the rail network. George Hudson was an East Yorkshire Tory, suddenly wealthy in his late twenties when he inherited £20,000. In 1833 he had been a joint founder and director of a new railway venture, the York and North Midland Railway, whose original purpose was to transport cheap coal from Leeds to York along a horse-drawn line. The Parliamentary Act for the 30-mile line was passed in 1835 and the line fully open by 1840. Like other ventures of the 1840s railway boom, it was unexpectedly high levels of passenger traffic which ensured that shareholders reaped 10 per cent dividends per annum for the best part of the decade.[13]

Hudson's domineering personality and political ambition created an atmosphere of autocratic suppression in the company. He was the very embodiment of the Communists' loathed capitalist tyrant. Men who dared to question his methods were threatened, or sacked. Hudson became responsible for almost all investment, expansion and operational decisions. His financial management, like that of his friend and collaborator George Stephenson, was inefficient, poorly documented and self-serving. But the profits which he was able to deliver to investors, his often shrewd acquisitions, and his clever involvement of senior government figures, masked his improprieties for the best part of ten years. He was able to plan and raise the capital for new lines more or less purely on his reputation for delivering profits. He built grandiose stations and magnificent bridges, was elected Member of Parliament for Sunderland, and no less a pen than that of Sydney Smith dubbed him 'The Railway King'. The railway boom of the 1840s was the ultimate expression of laissez-faire capitalism in all its glory and squalor.

George Hudson's fall, when it came in 1849, was swift. He had been fiddling company accounts for years, exaggerating profits, creaming off contracts, investing in new lines for personal aggrandizement rather than economic benefit, when a committee of enquiry began to uncover his systematic financial failures. Forced to resign his many chairmanships and pay investors back with his own money, he left England a bankrupt.

Founded on plans to move goods from one side of the country to the other and from north to south, the rail network was exploited by labourers, shopkeepers, clerks and the intellectual classes for their own ends. Many observers of this brave new world were forced to change their views. Some had been dead set against the railways but now came to accept them as either inevitable and necessary evils, or as a liberating force for good. Others, who had seen romance in the early locomotive and the stirring robustness of the engineers and their navvies, felt that the railways' industrial reality was coming a little too close for their liking. William Wordsworth, now Poet Laureate and very much an establishment voice, managed to execute a complete volte-face in the ten years between his elegiac 'Steamboats, Railways and Viaducts' of 1833 and the sonnet 'On the Projected Kendal and Windermere Railway' of 1844:

> Is then no nook of English ground secure
> From rash assault? Schemes of retirement
> In youth, and 'mid the busy world kept pure
> As when their earliest flowers of hope were blown,
> Must perish; – how can they this blight endure?
> And must he too the ruthless change bemoan
> Who scorns a false utilitarian lure
> 'Mid his paternal fields at random thrown?
> Baffle the threat, bright Scene, from Orresthead
> Given to the pausing traveller's rapturous glance:
> Plead for thy peace, thou beautiful romance
> Of nature; and, if human hearts be dead,
> Speak, passing winds; ye torrents, with your strong
> And constant voice, protest against the wrong.[14]

The Lakeland poet who had followed his heart to the streets of revolutionary Paris, who had declared that it was bliss in that dawn to be alive, had become a grumpy old man frothing in vain at the rape of progress in his own violated back garden, surviving, as Thackeray had wistfully said, like Father Noah out of the Ark.

There is no such impression of a guttering candle about the greatest painter of the day. J. M. W. Turner, who had watched and dispassionately portrayed the relationship between technology, society and land-scape since the beginning of the century, exhibited in this same year his famous *Rain, Steam and Speed*, a flashing and subtle commentary on the merging of past, present and future in a great roaring hiss of accelerated time.

This was the unstoppable tide of modernity, ridden not just by artists and poets, engineers, day-trippers and political interest groups, but by all sorts of opportunists.

Many years earlier an unknown banker had wagged a warning finger at Charles Babbage: 'Ah, I don't approve of this new mode of travelling. It will enable our clerks to plunder us, and then be off to Liverpool on their way to America at the rate of *twenty* miles an hour.' [15] He had been right. Criminals were only too happy to put 20 miles an hour between themselves and their pursuers.

John Tawell had precisely this idea in mind when, in 1845, he murdered his lover Sarah Hart. [16] Tawell was a native of Whitechapel, a convert in his teenage years to Quakerism. After marrying to legit-imize the baby of a girl he had seduced, he found work with a Quaker druggist in Cheapside and proved an adept pupil until he was found to have passed a ten pound note forged by himself. Condemned to death at his trial, Tawell had his sentence commuted to 14 years' transporta-tion after the intervention of his employers. Tawell did very well in Australia. He acquired a position as a druggist in a convict hospital, then as clerk to a member of the Sydney Academy. In 1820 his employer successfully petitioned for his pardon and he set up his own chemist's shop in Sydney, where his wife and child joined him three years later. They lived as Quakers and built up a substantial business exporting whalebone to the London market.

The Tawells and their two sons returned to England in 1831. Ill-health plagued them. By 1838 both sons and wife were dead and Tawell had embarked on an affair with the nurse he had employed to look after his wife during her last illness. He and Sarah Lawrence had two children but were never married. Instead, in 1841 Tawell married a Quaker widow who ran a school in Clerkenwell. Sarah Lawrence changed her surname to Hart and moved to a cottage in Slough, where

Tawell continued to visit her and pay maintenance for their children. Whether his wife knew of her and her children at the time was never revealed.

By 1844 Tawell was overwhelmed with debt. On the first day of the New Year, 1845, he caught a train from Paddington to Slough and, bearing a bottle of stout laced with prussic acid (cyanide) poisoned his lover. He left her writhing and groaning on the floor of her cottage and ran to Slough station to catch the 7.42 p.m. train to Paddington. Sarah's agonies attracted her neighbours, but they arrived too late to save her. The alarm having now been raised, the local vicar ran to the station to see if Tawell could be stopped. He got there just in time to see Tawell board the departing train and disappear eastwards.

Tawell ought surely to have known that Slough station was equipped with a Wheatstone and Cooke alphabetic telegraph: it was after all famous as the first such system in the world. It had been much improved since its first installation, the wires now suspended from posts alongside the railway tracks. The government had its own line connecting the Admiralty in Whitehall with the Royal Naval base at Portsmouth. The birth of Queen Victoria's second son had been announced from Windsor by telegraph. But Tawell was either bizarrely unaware of the telegraph or had not anticipated its rapid use. Alerted by the vicar, the stationmaster at Slough, a Mr Howell, sent the following message (at about a third of the speed of light) to Paddington, substituting the letter K for the letter Q which did not exist on the five-needle alphabetic telegraph:

> A murder has just been committed at Salt Hill and the suspected murderer was seen to take a first class ticket to London by the train that left Slough at 7.42pm. He is in the garb of a Kwaker with a brown great coat on which reaches his feet. He is in the last compartment of the second first-class carriage.[17]

At Paddington the Duty Sergeant of the Great Western Railway Police put a plain coat over his uniform and, easily recognizing Tawell from the description, followed him when he alighted from the train. The following day Tawell was arrested and, despite a desperate defence by his lawyers (the so-called 'apple-pip' defence) was condemned to death for a second time. He was hanged on 28 March 1845 on the balcony of the County Hall in Aylesbury, Buckinghamshire, the first criminal to have been apprehended by the use of the electric telegraph.

By the time John Tawell's capture brought the telegraph so spectacularly into the public mind, Charles Wheatstone was in bitter dispute with his one-time collaborator William Cooke. They had fallen out, as so many co-inventors do, over the division of rights to their patents and various improvements and applications. One witness called in Wheatstone's favour was his old friend John Martin, who happily testified to a long interest in Wheatstone's telegraphy experiments. Martin recalled very well the day in 1840 when Wheatstone had told him of his idea to synchronize clocks by use of distant telegraphic impulses. On grasping the implications of this idea Martin had said, 'You propose to lay on time through the streets of London as we now lay on water.'[18]

Martin, whose failing eyesight during the 1840s had put an end to his lucrative engraving career, was nevertheless showing few signs of artistic or intellectual decline. He was still painting watercolours and large oil canvases. He tried his hand at a grand fresco of *The Trial of Canute* as an entry in the competition to decorate the new Houses of Parliament, though he did not win. His enthusiasm for schemes to improve London had not waned either. He drew up plans for a circular underground railway and founded the Metropolitan Sewage Company to exploit his plans for clean water and sewage recycling. As always, he was supported by many disinterested public figures and by his friends Etty, Faraday and Turner. But even now, in the wake of Edwin Chadwick's shocking *Report into the Sanitary Conditions of the Labouring Population of Great Britain* of 1842, London was not ready to be saved from herself. It took another deadly cholera outbreak in 1848–9 and the Great Stink of the following decade before Martin's schemes were posthumously implemented by Joseph Bazalgette, a man whose name is revered in the annals of the city. Martin's fellow Geordie, John Snow, is equally revered by scientists as the doctor who established the crucial link between cholera and contaminated water. Martin himself is almost forgotten.

In the summer of 1845 John Martin went for an afternoon ramble with his son Leopold through the fields and villages of North London. Quite by chance they fell in with John's friend, the novelist William Harrison Ainsworth. At Ainsworth's house, Kensal Manor Lodge, they stopped to admire the magnificent view south across the Thames Valley. The silhouetted dome of St Paul's Cathedral dominated the skyline while a distant ribbon of silver betrayed the course of the Thames. To the east, close to Marylebone where the Martins had now lived for nearly thirty years, regular puffs of white steam rose from a locomotive pulling out of Euston Station on its way to Birmingham. There was something of a breeze: on a still day the view would be obscured by the smoke of more than a million fires.

The Martins were invited to enjoy a glass of wine in the writer's spacious and charming study and then, the weather remaining fine, Ainsworth persuaded John and Leopold to join him in a visit to the cemetery at Kensal Green to view the memorials of distinguished men and women they had known. These included their friends the poets Allan Cunningham and Tom Hood, and would shortly also include Sydney Smith. Harrison Ainsworth himself was to be buried here, as were many other Prometheans.

Martin and Ainsworth paused for a moment before a Graeco-Roman monstrosity which commemorated the circus horseman and entrepreneur Andrew Ducrow, a famous performer and manager who died in 1842. His immodest epitaph read, 'Erected by genius for the reception of its own remains'. Ainsworth and Martin might have raised their eyebrows at such a bold claim: between them they had known the Shelleys and the Brunels, Dickens, Babbage and Cruikshank, Faraday, the Stephensons, Turner and Constable, Melbourne and Disraeli.

Martin was still socially and politically active. He successfully petitioned Peel for a government position for his son Charles, later to become a respected portraitist. He still enjoyed long walks with Leopold across what little of London's countryside remained. They used often to climb Hangar Hill where Lady Byron lived. In earlier years they would encounter her daughter Ada, a 'timid, delicate but beautiful, child-like girl' out riding her pony.[19] The grown-up Ada, Countess Lovelace, was a woman of formidable intellect. Taught mathematics by her mother and by Mary Somerville, she forged passionate, if platonic, relationships with Charles Babbage, many of whose notes on the Analytical Engine she transposed and edited; with Michael Faraday, with Dickens and with Charles Wheatstone. She is said by some to have written the first computer program. Her untimely demise in 1852 at 36, the same age as her father, was caused by appalling medical negligence: she was bled to death.

In June 1846 Leopold Martin was asked by his father to take a letter and a sum of money to his very old friend, the artist Benjamin Robert Haydon. Haydon, three years older than John Martin, was 'a fine-looking, hawk-eyed, Roman-nosed bald-headed man … a wonderful talker'.[20] They had been introduced many years before by Charles Muss and had followed almost exactly parallel careers. Both were passionate, fierce self-advocates. Both had been rejected by the Royal Academy. Haydon was also, like Martin, a provincial, in his case a native of Plymouth (he went to the same school as Joshua Reynolds) arriving in 1804 in London, where he entered the Academy as a student. A technically much finer painter than Martin and capable of penetrating portraiture as well as grand historical designs, his most celebrated works were *The Judgement of Solomon* (1814), *Christ's Entry into Jerusalem* (1820) and the *Resurrection of Lazarus* (1823).

Like Martin, of whose imagination he was in awe but of whose technique he was privately critical, he fell into debt painting unfashionable historical subjects and dining too well. He was arrested three times and spent two periods in the King's Bench Prison. Unlike Martin he was a poor manager of patrons – he argued with them and failed to deliver on important commissions. On more than one occasion Martin lent him money and in the early 1830s the two fell out. This was during the period when Martin's own finances had become disastrous. When he asked Haydon to repay a loan, Haydon, apparently unaware that Martin was in similar straits, was furious. The friendship lasted, however: the two men were united by passion and a sense of rejection by the establishment. They had many friends in common, including

Henry Brougham and the Hunts, and they shared the radical political views of many of their artistic peers. Like Martin, Haydon also entered and failed to win the competition to paint frescoes for the new Parliament buildings. In Leopold Martin's view Haydon had never been understood by the public: 'His works were too large in every way.' [21]

Leopold was not very much amused when he arrived at the Haydon household (there was a wife and three surviving children) on the Edgware Road to find the place in a state of chaos but the table laden with fine food and wine. 'To my mind,' Leopold wrote, 'my father's benevolence was out of place.' But John Martin was sufficiently concerned by what Leopold had told him that he determined to go and see Haydon himself. The next day John and Leopold both walked over to the Haydons'

> … and to our utter dismay we found the house in the most frightful disorder. The poor wife was weeping, seemingly in deep anxiety and distress, her condition being truly both painful and shocking to witness. We gathered that in the morning, in front of his easel, with an unfinished canvas upon it, Haydon had been found a corpse. Disappointment, together with anxiety, had turned and quite crushed an overworn brain, and in a frenzy he had terminated his existence. [22]

Before cutting his throat and then shooting himself in his studio Haydon had written a short note quoting Lear: 'Stretch me no longer on this rough world.' [23]

John Martin had not given up on the world during his depression of the 1830s. Ten years on he believed he still had important work to complete. In 1848 the Martins moved house for the last time: to Lindsey House on the Thames in Chelsea, a short walk from Turner's shabby residence on Cremorne Road. Joseph Bramah and the Brunels had previously lived here and Whistler later occupied a house in the same complex of buildings, a magnificent remnant of a former palatial mansion of the seventeenth century. The views along the Thames were those that inspired Turner, and Martin arranged to be called by a neighbouring boatman if the night sky was worth coming out to paint. He and Turner together would watch the sun go down from a rowing boat.

Martin no longer needed the large printing establishment which he had built at Allsop Terrace because he could no longer engrave. Now, finally, he concentrated on painting a series of mature canvases which culminated in his *Judgement* trilogy. *A Port, Moonlight* (a watercolour of 1847) and *Arthur and Aigle in the Happy Valley* (a wonderfully romantic evening setting painted in 1849 and now exhibited at the Laing Art Gallery in Newcastle) have an intensely poetic quality. *The Last Man* (1849: Walker Art Gallery, Liverpool) was a post-apocalyptic message of acceptance, already treated by Thomas Campbell in an 1823 poem and by Mary Shelley in her thoroughly depressing novel of 1826.

But if his public thought that John Martin had done with the apocalyptic sublime, they were wrong. *The Destruction of Sodom and Gomorrah* (1852: Laing Art Gallery) is Martin's best-known surviving work. It is a furnace of yellows and reds, riven by a signature bolt of lightning with the pathetic figures of Lot, his daughter and fossilized wife dumbstruck in the foreground. Visitors still experience a disconcerting feeling of physical heat emanating from the canvas. Like many of Martin's works, its influence survives in unexpected places.

It is entirely reasonable that directors of cinematic epics like D. W. Griffiths and Cecil B. de Mille should have looked to John Martin for inspiration. One can see his explosive sense of scale and awe in the work of Ridley Scott (a fellow Geordie), of Steven Spielberg and George Lucas. Conan Doyle, Jules Verne, H. G. Wells, Victor Hugo and Rider Haggard were all influenced by his epic sense of space and dramatic narrative structure.[24] Dante Gabriel Rossetti was a lifelong admirer. Oddly, or perhaps not so oddly, Martin has also recently become an icon for a generation of thrash-metal rockers and their concept-album cover artists. He would have been delighted.

The *Judgement* trilogy, his last great series, was painted between 1851 and 1853, during which time Martin had taken to spending long vacations in the Isle of Man (he is buried there in Kirk Braddon churchyard). The first of the trilogy, *The Last Judgement* (1851: Tate Britain) is a rather unsatisfactory composition of angels, penitents and lowering skies. It has all the recognizable Martinian features but none of the pugnacious bravado of *Sodom and Gomorrah* or the epic early works. The same can be said of the third painting in the trilogy, *The Plains of Heaven* (1853: Tate Britain). But the centrepiece of the trilogy, the magnificent *Great Day of His Wrath* (1852–3: Tate Britain) stands as a final triumph of Martin's unquenchable spirit. It is an angry red cine-matic whip-pan of pyrotechnic brilliance by the master of the genre. He has found the means to fold the entire earth in on itself in a last seismic retributive act of God's wrath. Temples, citadels, palaces and one of Brunel's locomotives topple into an abyss with numberless infinities of human weakness, folly and sin. Martin's grand finale is not, for once, a portrayal of retribution on the individual. This time, at last, it is the entire human race which must pay for its collective acts of hubris. Nothing he ever painted more perfectly conveys the extraordinary conception of John Martin's Promethean imagination. One day a publisher will think to use it as the front cover for an edition of the Communist Manifesto.

John Martin died in the Isle of Man after a stroke, in 1854. And what of his last surviving brother, William? Through the 1830s and 1840s, his wife having died in 1832, he remained for the most part in Newcastle, interesting himself in everything, writing shocking doggerel, innumerable tracts and pamphlets and ranting against false philoso-phers, learned humbugs and all those unscrupulous men of fact or fiction who had stolen his ideas. Many years after William's death the artist William Bell Scott [25] recalled how, in the late 1840s, he had met the famous artist's brother:

> … a third brother, quite as mad as the incendiary but more innocent. He was habitually to be met in the principal thorough-fares, generally with a pamphlet in his hand, which he was willing to dispose of … A few weeks later we encountered the well-known figure in his extraordinary skull-cap decorated with tortoise-shell, and a military surtout closely buttoned to the throat. Captain Weatherley [James Dent Weatherley who served under Wellington

during the Peninsular War and later became Mayor of Newcastle]
as his manner was, received him in the friendliest way, and listened
to the information that Martin's claim to be the inventor of the
High-Level Bridge then building over the Tyne – a railway scheme
designed, if I remember right, by Stephenson the younger – was now
in print, and would be forwarded tomorrow. He then introduced me
as a great London artist come to educate the people of the north,
when Martin, with an exaggerated politeness, drew his feet
together, bent forward, lifted his tortoise-shell hat high in the air,
and answered, 'Gratified to meet you, Sir! I am the Philosophical
Conqueror of all nations, that is what I am! And this is my badge,'
at the same time unbuttoning his surtout he showed a medal as
large as a saucer, which was hung round his neck by a ribbon. It
was not a medal at all, and he was manifestly crazed, yet he had that
about him that made one treat him with respect. A noble presence
even was his, although he was poor enough to sell his pamphlets
thus on the street, which pamphlets were of course only evidence
of his craze. Shortly afterwards he disappeared. I understand his
brother the painter had taken care of him by carrying him to
London.[26]

William Martin died in his famous brother's home at Lindsey House,
Chelsea, on 9 February 1851. He missed, by three months, the opening
of the Great Exhibition, where Joseph Bramah's unpickable lock was
finally picked 67 years after it was first displayed in his shop window. An
American locksmith called Alfred Hobbs spent more than 40 hours
over 10 days in opening it and duly claimed the 200-guinea prize.

Notes and references

CHAPTER ONE The chains of slavery

1 *Newcastle Courant*, Saturday 23 November 1771.

2 *Newcastle Courant*, Saturday 30 November 1771.

3 *Newcastle Courant*, Saturday 30 November 1771.

4 Edmund Burke, *Reflections on the Revolution in France*, 1790.

5 *Newcastle Courant*, Saturday 30 November 1771.

6 *Newcastle Courant*, Saturday 30 November 1771.

7 Gottschalk (1927), p. 25.

8 Jean-Paul Marat (1774), *The Chains of Slavery*, p. 3.

9 Jean-Jacques Rousseau (1750), *Discourse on the Arts and Sciences*, second part.

10 A 'keel' was an ancient craft derived from an original Anglo-Saxon design; with a steering oar that doubled as a paddle, a keel held about 21 tons, or 20 chaldrons, of coal. Keelmen were closely regulated, militant and jealously protective of their trade.

11 Quoted by Briggs (1979), p. 8.

12 Jean-Jaques Rousseau (1762), *Du contrat social*.

13 It seems they later went through a more formal service closer to home, for they appear in the marriage register of a church near Hexham the following year. Information supplied by Henry Swaddle.

14 William Martin (1833), *A Short Outline of the Philosopher's Life*.

15 Jonathan Martin (1829), *The Life of Jonathan Martin the Incendiary of York Minster*, 6th edn.

16 Franklin was portrayed in a very obvious Promethean allegory by J. H. Fragonard in 1778: *Au génie du Franklin*, Bibliothèque Nationale, aquatint.

17 Karl Marx (1841), *Dissertation and Preliminary Notes on the Difference between Democritus' and Epicurus' Philosophy of Nature*, University of Jena.

18 Jonathan Martin (1829), *The Life of Jonathan Martin the Incendiary of York Minster*, 6th edn.

CHAPTER TWO The rights of man

1 Singer (ed.). (1958), *History of Technology*, pp. 106–7.

2 Quoted in Smiles (1863): *Industrial Biography*.

3 Adam Smith (1776), *An Inquiry in the Nature and Causes of the Wealth of Nations*, Book 1, chapter 8. The italics are Smith's.

4 Thomas Jefferson's last letter, quoted by Williams (1971), p. 4.

5 Thomas Paine (1776), *Common Sense*.

6 Derry and Williams (1961), p. 450. Paine's bridge, of which he produced a model, had 13 iron ribs to celebrate the liberation of the 13 American colonies.

7 Unitarians believed in a single indivisible God; open denial of the Trinity was outlawed in Britain until 1813.

8 William Wordsworth, 'The French Revolution, as it appeared to Enthusiasts at its Commencement', *The Prelude*, Book XI, pp. 108–12.

9 Gottschalk (1927), p. 39, quoting from *l'Offrande de la Patrie* in 1789.

10 Jean-Paul Marat, *C'en est fait de nous*, 26 July 1790.

11 Henry Fuseli, *Prometheus* (1770–71), pencil and brown ink on paper; *Hephaestus, Bia and Crato securing Prometheus on Mount Caucasus* (1800–10), pencil and grey wash on paper. John Flaxman, *Prometheus Bound* (1794), pen, ink and pencil on paper. Richard Cosway, *Prometheus* (1790–1800), pen and brown ink on paper. George Romney, *Prometheus Bound* (1779–80), chalk on laid paper. William Blake, *Los and Orc* (1790–94), pen, ink and watercolour on paper. From the Tate Gallery, *Gothic Nightmares* exhibition catalogue, edited by Martin Myrone (2006). For Blake's status as a Promethean see Donoghue (1973), p. 62. It is also possible that Blake saw himself not as Prometheus, but as the receiver of the illicit flame.

12 Marat, quoted by Gottschalk (1927), p. 131.

13 Mary Wollstonecraft, *A Vindication of the Rights of Woman* (1792), p. 93.

14 *The Death of Marat* (1793), Musées Royaux des Beaux-Arts at Brussels.

15 Paine himself told the story eight years later. However, in the second edition of *The Age of Reason*, Preface to Part II, he wrote that he believed his illness in the Luxembourg prison was the means of his avoiding execution, though by what means he was not sure. Aldridge (1960), p. 217.

16 Thomas Hardy (1799), *An Account of the Origin of the London Corresponding Society.*

17 Woodcock (1946), p. 100.

18 Woodcock (1946), p. 16

CHAPTER THREE Children of the Revolution

1 Celia Brunel Noble (1938), *The Brunels, Father and Son*, p. 5.

2 Noble (1938), p. 19.

3 Pierre-Simon de Laplace (1749–1827). His *Traité de mécanique Céleste* was published in 1799. Somerville was commissioned to translate it by Henry Brougham for the Society for the Diffusion of Useful Knowledge. *The Mechanism of the Heavens* appeared in 1831, published by John Murray.

4 Maria Edgeworth to Mrs Buxton, 17 January 1822.

5 Mathias (1983), p. 167.

6 Thomas Malthus, *Essay on the Principle of Population* (1798).

7 Paul Webb, in *The British Navy and the Use of Naval Power in the Eighteenth Century*, ed. Jeremy Black and Philip Woodfine (1988), p. 207.

8 This was not an official title. The first to hold the title as an office was Henry Campbell-Bannerman in the twentieth century; eighteenth-century

prime ministers held the office of First Lord of the Treasury, as prime ministers do today.

9 Clarke (1977), *The Price of Progress*, p. 13.

10 J. Sykes (1807), *Local Records.*

11 The staithes were depicted in vignettes of the time by Newcastle engraver Thomas Bewick; by J. M. W. Turner on his voyage north in 1822; and later by the great watercolourist of the coal-fields, Thomas Hair.

12 J. Sykes (1794), *Local Records.*

13 William Martin (1833), *A Short Outline of the Philosopher's Life* (1833), p. 17. Iron rails had first been used at Coalbrookdale in 1767; but they were not generally employed until after about 1810 because of the brittleness of cast-iron.

14 William Martin (1833), *A Short Outline of the Philosopher's Life*, p. 17. A currier is a treater of tanned leather.

15 The standing jump was an Olympic event from 1900 to 1912. The last world record was that of the American athlete Ray Ewry, set in 1904, of 11 feet 5 inches. Ewry remains one of the most successful Olympic athletes ever. Having recovered from childhood polio, he won ten titles between 1904 and 1908.

16 Adamson (1877), p. 32.

17 Clarke (1977), *The Price of Progress*, p. 128.

18 Hewitson (1999), p. 11.

19 A fugleman was a sort of drill-sergeant.

20 Hartley (1966), p. 12. The debate over the composition of air had in the 1770s largely been resolved by Priestley, Scheele and Lavoisier after a long and sometimes bitter dispute.

21 From Humphry Davy, *The Sons of Genius* (1799), published by Robert Southey.

22 Hartley (1966), p. 24.

23 Pearce Williams (1965), pp. 63ff.

24 It was given him by the biographer J. A. Paris in his *Life of Sir Humphry Davy* (1831).

CHAPTER FOUR Mechanics of war

1 J. Sykes (1796), *Local Records.*

2 J. Sykes (1796), *Local Records.*

3 Padfield (2003), p. 117.

4 Quoted by Harris (1981), p. 112.

5 Singer (ed.). (1958), p. 423.

6 John Farey, quoted by Singer (ed.). (1958), p. 423.

7 Smiles (1863), p. 176.

8 *Remarks on the Introduction of the Slide Principle in Tools and Machines employed in the Production of Machinery*, in Buchanan's *Practical Essays on Mill Work and other Machinery*, 3rd edn, p. 397.

9 Singer (ed.). (1958), p. 426.

10 *Falconer's Marine Dictionary*, abridged edition by Claude S. Gill (1930), p. 20.

11 Marc Brunel, quoted by Noble (1938), p. 21.

12 Smiles (1863), p. 188.

13 James Nasmyth, *Autobiography* (1885), chapter 7.

CHAPTER FIVE Brothers in arms

1 Adamson (1877), p. 38.

2 Balston (1945), p. 119

3 John Martin, autobiographical notes in *The Athenaeum*, Saturday 14 June 1834.

4 John Martin, autobiographical notes in *The Illustrated London News*, Saturday 17 March 1849.

5 John Martin, autobiographical notes in *The Illustrated London News*, Saturday 17 March 1849.

6 Jonathan Martin: *The Life of Jonathan Martin the Incendiary of York Minster*, 6th edn, 1829. Jonathan's English was poor: the text of his life, hanging participles included, was edited by a more literate man.

7 HMS *Hercules* was a French 74-gun two-decker captured by HMS *Mars* off the Bec du Raz in 1798. She was put into ordinary at Chatham after the siege of Copenhagen 1807 and broken up at Portsmouth in 1810.

8 Cuthbert Collingwood was born in Newcastle in 1748. The oldest of three brothers, he could not follow his father into the trading business because his father had gone bankrupt. In his case, the sea was an obvious choice of career: it required no capital and promised both adventure and prize money.

9 Admirers on both sides of the English Channel identified Napoleon with Prometheus; among them Byron, whose 1814 'Ode to Napoleon Bonaparte' laments his downfall. Bonaparte himeslf struck a campaign medal in Italy portraying Prometheus on the reverse.

10 Jonathan Martin 1829, *The Life of Jonathan Martin the Incendiary of York Minster*, 6th edn.

11 Leman Thomas Rede, author of *York Castle in the Nineteenth Century* (1831).

12 John Martin, autobiographical notes in *The Illustrated London News*, Saturday 17 March 1849.

13 Charles Musso took the surname Muss when he moved to London in 1806, perhaps for the same Anglicizing reasons that immigrants chose during the war years of the twentieth century.

CHAPTER SIX A million fires

1 Sir Richard Phillips: *A Morning's Walk from London to Kew* was first published in 1817; but

internal evidence shows that his walks must have begun while the Peninsular War was still being fought between 1808 and 1812.

2 Phillips (1820), p. 6.

3 Phillips (1820), p. 12.

4 Phillips (1820), p. 49.

5 William Blake: 'London', from *Songs of Experience* (1791).

6 Phillips (1820), pp. 130–1.

7 According to Leopold Martin: Reminiscences, *Newcastle Weekly Chronicle*, 16 March 1889.

8 According to Leopold Martin: Reminiscences, *Newcastle Weekly Chronicle*, 16 March 1889.

9 James Hamilton (1998), *Turner and the Scientists*, London, Tate Publishing, p. 10.

10 The full title of the work was *The Battle of Trafalgar, as seen from the Mizzen Starboard Shrouds of the Victory*, exhibited in Turner's Harley Street gallery in 1806.

11 John Martin, autobiographical notes in *The Illustrated London News*, Saturday 17 March 1849.

12 Klingender (1972), p. 130.

13 Sandemanians were a Christian sect founded in the early part of the eighteenth century by the Scot Robert Glas. They denied the Trinity and adhered to strict Congregational rules. William Godwin had been brought up as a Sandemanian and Thomas Spence espoused their belief that land should not be held as personal property.

14 *An Experimental Enquiry concerning the Source of the Heat which is Excited by Friction* (1798).

15 *The Literary Tablet*, London, 24 October 1805.

16 Pearce Williams (1965), p. 19.

17 Sir Humphry Davy: *A Discourse, Introductory to a Course of Lectures on Chemistry, Delivered in the Theatre of the Royal Institution on the 21st of January, 1802* (London: Sold at the House of the Royal Institution, 1802).

18 See above, chapter 3, note 24.

19 Lawson Cockcroft (2001), in an article on Accum posted on the Royal Society of Chemistry's website; there is no fully published biography of Accum.

20 Balston (1945), p. 119.

21 Balston (1945), p. 120.

22 Balston (1945), p. 141.

23 Invented by a Scot, the Reverend Robert Stirling in 1816, the heat economizer used temperature differentials to recycle exhaust gases and increase the efficiency of many engines; its use has been revived in modern condensing gas boilers.

24 William Martin (1833), *A Short Outline of the Philosopher's Life*, Newcastle upon Tyne.

25 Leopold Martin: 'Reminiscences of John Martin', *Newcastle Weekly Chronicle*, 5 January 1889.

26 The pamphlet was published by Philadelphia Phillips at Worthing in February 1811, and circulated anonymously by Shelley and Hogg from University College. The two were expelled before the end of their first year, in March 1811.

27 Thomas Jefferson Hogg (1858), *The Life of Percy Bysshe Shelley*, vol. I, pp. 69–70.

28 Pendered (1923), p. 56.

29 Spencer Perceval was assassinated by a bankrupt Liverpool banker, John Bellingham, on 12 May 1812 in the lobby of the House of Commons.

30 Quoted by H. Jennings (1985), *Pandaemonium*, pp. 131–2.

31 William Feaver (1975) traces it to *Tales of the Genii* (1762) by James Ridley. The Tales are a pastiche of the *Arabian Nights*.

32 There is also a strong argument for equating some aspects of the God/Adam relationship in *Paradise Lost* with that of Prometheus/Zeus. However, Denis Donoghue, in arguing that Milton should be regarded as part of a group of 'Promethean' writers, warns of overinterpreting this aspect of *Paradise Lost*. See Donoghue (1973), p. 58.

33 See below, chapter 8.

34 Shelley was to make an explicit comparison in the preface to his *Prometheus Unbound* of 1820; see below, chapter 9.

35 Brougham was pronounced 'Broom' except by his enemies, who referred to him variously as Bruffam, The Learned Friend (Thomas Love Peacock in Crotchet Castle, 1831) Beelzebub, the Arch-fiend (Thomas Creevey) and so on.

36 Treasure (1997), p. 314.

37 Klingender (1972), p. 29.

38 Edmund Blunden, *Leigh Hunt's 'Examiner' Examined* (London: Harper & Brothers, 1931), p. 23.

39 Roe (2005), p. 361.

CHAPTER SEVEN Peace dividend

1 Knight (1992), p. 69.

2 Dance was a founder of the Royal Philharmonic Society.

3 Pearce Williams (1965), p. 29.

4 Guy (2003), p. 32.

5 *Durham County Advertiser*, 5 August 1815.

6 Guy (2003), p. 40.

7 The phenomenon that allows the human brain to join contiguous images together to create the illusion of a moving picture.

8 Noble (1938), p. 30.

9 Phillips (1820), p. 48. In 1819 Phillips met with Henry 'Orator' Hunt and Samuel Bamford, both on bail for their part in the so-called Peterloo Massacre. Phillips helped Bamford to draw up a petition to Parliament.

10 Byron's 'Ode to Napoleon' of 1814 reflects not only the poet's ambivalent love for the tyrant, but also a love of the Greek hero whose plays he had been forced to read and reread at Harrow (Clubbe, 1997).

11 Hamilton (1998), p. 92.

12 Utilitarianism is a theory which expounds the principle that goodness is whatever brings the greatest happiness to the greatest number of people. It was originally propounded by Jeremy Bentham (1748–1832), whose disciples included James Mill, the father of John Stuart Mill. Jeremy Bentham's brother, Samuel, was the Inspector-General of the navy dockyard who commissioned Marc Brunel's block-making scheme.

13 New Lanark, which finally ceased production in 1968, has now been sympathetically restored and has been designated a World Heritage site. It lies 25 miles southeast of Glasgow in a leafy gorge where the Falls of Clyde provided its motive power. The social housing scheme still operates.

14 Richardson (1961), p. 17.

15 Quoted by Fraser (1996), p. 223.

16 Quoted by Balston (1945), p. 37.

17 Leopold Martin: Reminiscences, *Newcastle Weekly Chronicle*, 17 April 1889.

18 C. R. Leslie (1860), *Autobiographical Recollections*, vol. I, p. 202. The beautiful grey sea painting in question was *Helvoetsluys*, exhibited in 1832.

19 The Department of Culture, Media and Sport granted an export licence for *Joshua* in 2003–4, considering that it was not sufficiently important to meet the third Waverley criterion which would have kept it in Great Britain, despite testimony to the contrary from the Senior Curator at the Tate Gallery. *Joshua* was sold for more than three million pounds.

20 Charles Lamb (1833), *The Barrenness of the Imaginative Faculty in the Production of Modern Art*.

21 John Martin, autobiographical notes in *The Illustrated London News*, Saturday 17 March 1849. This may also be a reference to the interpretation of *Joshua* as a celebration of Wellington's victory at Waterloo the previous year.

22 Quoted by Richardson (1961), p. 47. The

National Debt in 1815 at the end of the war was £834,000,000 according to Lane (1978), *Success in British History*, p. 74.

23 Lane (1978), p. 77.

24 *The Examiner*, 23 February 1817.

25 *The Examiner*, 9 February 1817.

26 *The Examiner*, 9 March 1817.

27 Thomas Gray (1782), 'The Bard: A Pindaric Ode'. The 'prophet of fire' is Elijah, whose passage to heaven in a chariot of fire precedes the coming of the Messiah. William Blake had painted the same theme in 1809.

28 Special jurors, selected by the Crown for sensitive or politically motivated cases, comprised men of property or wealth: bankers, merchants and esquires.

CHAPTER EIGHT A light in the darkness

1 Rev. John Hodgson (1813), *The Funeral Sermon of the Felling Colliery Sufferers*, Newcastle.

2 Rev. John Hodgson (1813), *The Funeral Sermon of the Felling Colliery Sufferers*, Newcastle.

3 R. L. Galloway (1882), *A History of Coal Mining in Great Britain*, Macmillan, p. 158.

4 Quoted by Galloway (1882), p. 160.

5 Galloway (1882), p. 144.

6 *The Examiner*, 24 October 1813.

7 Davy, *Works*, vol. 1, pp. 173–5.

8 Galloway (1882), p. 163.

9 Watson (1999). Watson shows that the published version of Davy's address to the Royal Society differs from the minutes taken at the time in including details of the wire gauze arrangement.

10 Watson (1999), p. 140.

11 Galloway (1882), p. 175.

12 Galloway (1882), p. 178.

13 Mary Shelley, *Journals*, p. 70, quoted by Dr Siv Jansson (1999), Introduction to *Frankenstein*, Wordsworth Classics.

14 Mary Wollstonecraft (1792), *A Vindication of the Rights of Woman*, p. 93.

15 *The Revolt of Islam* (1817), Canto V, 23.

16 Anonymous review in *The British Critic*, 9 April 1818, pp. 432–8.

17 The *Quarterly Review*, 18 (January 1818): 379–85.

18 *The Belle Assemblée, or Bell's Court and Fashionable Magazine*, 17 March 1818, pp. 139–42.

19 *Blackwood's Edinburgh Magazine* 2 (20 March/1 April 1818), pp. 613–20.

20 *Blackwood's Edinburgh Magazine* 2 (20 March/1 April 1818), pp. 613–20.

21 Jonathan Martin (1829), *The Life of Jonathan Martin the Incendiary of York Minster*, 6th edn, p. 19.

22 The existence of the *Steam Elephant*, an important stage in the development of the locomotive, has only been demonstrated in the last few years after detailed research. An exact full-scale replica now runs at the North of England Open Air Museum at Beamish, County Durham.

23 *Shields Gazette*, 25 August 2006.

24 Staithe is an old Norse word for a landing stage.

25 Balston (1945), p. 125, quoting from William Martin's autobiography.

26 William Martin, *Tracts* 29, Literary and Philosophical Society of Newcastle upon Tyne.

27 Galloway (1882), p. 179.

CHAPTER NINE *Belshazzar's Feast*

1 *Life at Fonthill 1807–22*, from the correspondence of William Beckford, transcribed and edited by Boyd Alexander. Rupert Hart-Davis (1957), p. 279: Friday 5 February 1819.

2 Quoted by Balston (1947), p. 48.

3 Martin to Ralph Thomas, according to Pendered (1923), p. 89.

4 London was known as Babylon-on-Thames, just as Washington DC today is referred to as Sodom-on-the-Potomac. For example, B. R. Haydon calls London 'our Babylon' in his diary (quoted by Jennings, 1985, p. 125). Samuel Bamford (1844) called the capital 'Babylon' in his *Passages in the Life of a Radical* (p. 29 in the Fitzroy edition). In Shelley's *Peter Bell the Third*, 'London is a city much like Hell' (Part 3, 1); and in 1824 Thomas Carlyle called London 'this Babel of a place'; quoted by Jennings (1985), p. 165.

5 From the Hutchinson (1929) edition of Shelley's *Complete Works*.

6 *The Examiner*, 3 January 1819.

7 Thackeray (1855); Pocket Classics edition (1995), p. 95.

8 Samuel Bamford (1844), *Passages in the Life of a Radical*; Fitzroy edition (1967), p. 135.

9 Samuel Bamford (1844), *Passages in the Life of a Radical*; Fitzroy edition (1967), p. 152.

10 Quoted by Humphrey Jennings (1985), *Pandaemonium*, p. 152.

11 'Sonnet: England in 1819' was published posthumously by Mary Shelley in 1839; from the Hutchinson (1929) edition of Shelley's *Complete Works*.

12 *Prometheus Bound* survives intact, but the second and third parts of Aeschylus' intended trilogy, *Prometheus Unbound* and *Prometheus Pyrphoros*, exist only as fragments.

13 Thomas Hutchinson (ed.) (1929), *The Complete Poetical Works of Percy Bysshe Shelley*, p. 267.

14 *Prometheus Unbound*, Act II, 71–2.

15 *Prometheus Unbound*, Act II, 99.

16 *Prometheus Unbound*, Act I, 550–3.

17 *Prometheus Unbound*, Act I, 560.

18 *Prometheus Unbound*, Act IV, 153–8.

19 *Newcastle Weekly Chronicle*, 2 March 1889.

20 Sydney Smith (1771–1844). Born in Woodford, Essex; buried, like so many Prometheans, in Kensal Green cemetery.

21 Letter to Lord Grey (1819), quoted by Taylor and Hankinson (1996), p. 127.

22 H. Viztelly (1893), *Glances back through Seventy Years*, p. 254.

23 Hugh Trevor-Roper (1983), 'The Invention of Tradition: The Highland Tradition of Scotland', in E. Hobsbawn and T. Ranger (eds) (1983), *The Invention of Tradition*, Cambridge, Cambridge University Press.

24 John was unlucky. Around the time of *Macbeth*'s exhibition a cartoon appeared depicting George IV as Macbeth with Lords Liverpool and Sidmouth, and Castlereagh as the three witches.

25 *Newcastle Weekly Chronicle*, 26 January 1889.

26 Pendered (1923), p. 233.

27 Treasure (1997), p. 358.

28 Quoted by Fraser (1996), p. 363.

29 Quoted by Balston (1947), p. 54.

30 Quoted by Balston (1947), p. 55.

31 There are two surviving oil-on-canvas versions of the *Feast* by Martin. One, in the Laing Art Gallery in Newcastle, is large at 8 ft by 5 ft, but is inferior in every respect to that possessed by the Yale Center for British Art which is smaller at 37 by 47 inches. Which canvas was that painted in 1821 is a moot point. The 'American' version certainly possesses the 'foxy hues' criticized by Ruskin. It is hard to imagine the Laing version creating the stupendous public impact recorded by contemporaries. It is possible that it is a copy by an artist other than Martin. Ian Balston (1947, p. 61), Martin's biographer, was of the opinion that both these versions are later copies.

32 *The Examiner*, 23 July 1820.

33 Noble (1938), p. 33.

34 Noble (1938), p. 44.

35 Institute of Electrical Engineering, *Chemical Lectures* 4, p. 506, quoted by Pearce Williams (1965), p. 89.

36 This much-used phrase, that scientists are like dwarves sitting on the shoulders of giants, was employed famously by Isaac Newton and before him Robert Hooke, but probably owes its origins to John of Salisbury who may himself have heard it from Bernard of Chartres in the twelfth century. See Kenneth Clark's *Civilisation*, p. 89.

37 Pearce Williams (1965), p. 154.

38 L. Cockcroft (2001), *Friedrich Carl Accum*, Royal Chemistry Society, p. 7.

CHAPTER TEN *Paradise Lost*

1 John Milton, *Paradise Lost* I: 651–4.

2 It is more properly called the War of Three Nations, since it began in Scotland and accounted for the lives of four hundred thousand Irishmen and women.

3 Klingender (1972), pp. 58ff. provides an invaluable account of these processes.

4 Klingender (1972), p. 58.

5 Balston (1947), p. 95.

6 Hamilton (1998), pp. 92ff.

7 The *Strickland Manuscript*, now in the possession of the North of England Open Air Museum at Beamish, is a great work of technical illustrative art. Strickland's conclusion, that the locomotive was the transport of the future, was initially suppressed by his employers, the Philadelphia merchants, who had favoured canals as the means of conquering America. That Strickland was given virtually unlimited access to Britain's technical secrets now seems extraordinary.

8 Jonathan Martin (1829), *The Life of Jonathan Martin the Incendiary of York Minster*, 6th edn, p. 25. Freestone is a fine-grained oolitic limestone or sandstone used for architectural carving. The term freemason is derived from the guild of freestone carvers. Jonathan was barely literate judging from his scribblings: the autobiography was probably dictated.

9 Jonathan Martin (1829), *The Life of Jonathan Martin the Incendiary of York Minster*, 6th edn, p. 27.

10 William Hazlitt (1825), 'Mr Coleridge', in *The Spirit of the Age*.

11 William Hazlitt (1825), 'Mr Coleridge', in *The Spirit of the Age*.

12 Hamilton (1998), p. 98.

13 H. Ward (1926), p. 9.

14 Society for the Diffusion of Useful Knowledge, *Prospectus*, 1829.

15 Quoted by James Second (2000), in *Victorian Sensation*, p. 46, n. 11.

16 *Crotchet Castle* (1831), chapter 17.

17 Leopold Martin: 'Reminiscences of John Martin', *Newcastle Weekly Chronicle*, 9 March 1889.

18 Leopold Martin: 'Reminiscences of John Martin', *Newcastle Weekly Chronicle*, 9 March 1889.

19 An 1823 oil sketch for the painting is in the possession of the Laing Art Gallery in Newcastle.

20 HL was Henry Leigh, Leigh Hunt's son. Leigh remained in Italy until the end of 1825.

21 *The Deluge* (1826) has been lost. The version now possessed by the Yale Center for British Art collection is that of 1834. See chapter 13.

22 'Christopher North' quoted by Balston (1847), p. 89. North was the nom de plume of John Wilson, the professor of moral philosophy at Edinburgh University.

23 Leopold Martin: 'Reminiscences of John Martin', *Newcastle Weekly Chronicle*, 26 January 1889.

24 According to Marc, quoted by Noble (1938), p. 61.

25 James Hamilton (1998, pp. 98ff) makes a strong case for ascribing this picture to the event of the banquet.

26 Quoted by Balston (1947), p. 120.

27 Quoted by Balston (1947), p. 107.

CHAPTER ELEVEN Playing with fire

1 Rolt (1960), p. 68.

2 Leopold Martin: 'Reminiscences of John Martin', *Newcastle Weekly Chronicle*, 30 March 1889. From what Leopold says in another article in this series (29 April 1889) John Martin, a keen student of all new engineering ideas, may have been present at this original bill hearing.

3 Reported by Leopold Martin: 'Reminiscences of John Martin', *Newcastle Weekly Chronicle*, 29 April 1889.

4 Fanny Kemble to a friend, 26 August 1830, quoted by Jennings (1985), p. 172.

5 Quoted by Rolt (1960), p. 158.

6 Quoted by Balston (1945), p. 56.

7 *Edinburgh Review*, June 1829.

8 *Edinburgh Review*, June 1829.

9 Morrison and Morrison (1961), p. xiv.

10 Hyman (1982), *Charles Babbage, Pioneer of the Computer*, pp. 121ff.

11 Babbage (1830), p. 207.

12 Balston (1847), pp. 130–1. John Martin's pamphlet *Outlines of Several Inventions for Maritime and Inland Purposes* was published in 1829.

13 Transcribed in *The Life of Jonathan Martin the Incendiary of York Minster*, 6th edn (1829), p. 39.

14 This was the testimony given by Jonathan

himself at his trial, taken down by a stenographer and therefore correctly spelt. *The Life of Jonathan Martin the Incendiary of York Minster*, 6th edn (1829), p. 48.

15 Quoted by Richardson (1891), p. 45.

16 According to Balston (1945), p. 69.

17 Balston (1945), pp. 78–9.

18 'On the Burning of York Minster', by William Martin. The poem is contained in *A Short Outline of the Philosopher's Life*. Newcastle, 1833. Richard himself was shortly to publish a long poem of over two thousand lines, *The Last Days of the Antediluvian World*, with a frontispiece by John.

19 There were four prime ministers in less than two years: Liverpool was briefly succeeded by Canning and Goderich before Wellington formed his own short-lived government. The Tories would finally be turfed out of office in 1830 after 23 years.

20 Quoted by Treasure (1997), p. 318.

21 *The Life of Jonathan Martin the Incendiary of York Minster*, 6th edn (1829), p. 40.

22 Henry Brougham was prominent in moves to clarify the law on criminal insanity after the case of Daniel McNaughton who, in attempting to assassinate the Prime Minister Sir Robert Peel in 1843 managed only to kill Peel's Private Secretary. Jonathan Martin's case was used as an important precedent for the judgment in that case. See below, chapter 15.

23 *The Life of Jonathan Martin the Incendiary of York Minster*, 6th edn (1829), pp. 41–2.

24 The hospital at that time was situated in St George's Fields, Lambeth, in the buildings which now house the Imperial War Museum.

25 Sir Roger Ormrod: 'The McNaughton Case and its Predecessors', in West and Walk (1977), pp. 5–6.

26 The pounds of iron refer to his manacles, attached by chains to an iron belt.

27 Quoted by Balston (1945), pp. 99–100.

CHAPTER TWELVE The 'democratical' principle

1 Wells (1997), p. 36.

2 Morrison and Morrison (1961), p. 113.

3 Jennings (1985), p. 175.

4 Quoted by Jennings (1985), p. 179.

5 Wellington, 2 November 1830, quoted by Lane (1978), p. 113.

6 Richardson (1961), p. 107.

7 Richardson (1961), p. 174.

8 Cobbett, quoted by Wells (1997), p. 42.

9 Acland (1948), p. 52.

10 Cecil (1954), p. 224.

11 Disraeli's first novel, *Vivian Grey*, was published in 1826 to popular acclaim and critical derision.

12 Cecil (1954), p. 226.

13 Kegan Paul (1876), p. 321.

14 *Newcastle Weekly Chronicle*, 19 January 1889.

15 Charles Macfarlane (1917), *Réminiscences of a Literary Life*, pp. 98f.

16 William Godwin, *An Enquiry concerning Political Justice*, Book IV, chapter IV. Godwin argues both sides of the case with equal elegance and conviction.

17 Macfarlane (1917), pp. 98f.

18 *Newcastle Weekly Chronicle*, 30 March 1889.

19 Quoted by Hamilton (2002), p. 247.

20 Dinwiddy (1986), pp. 49ff.

21 Quoted by Brock (1973), p. 191.

22 It was never discovered if there was an actual individual responsible for fomenting the riots, machine-breaking and arson. Probably there was no Captain Swing, though many candidates were suggested, including Cobbett.

23 Quoted by Brock (1973), p. 214.

CHAPTER THIRTEEN Babylon-on-Thames

1 Raumer (1836), I, p. 7.

2 Pückler-Muskau (1832), vol. 3, p. 52. The German was translated by Sarah Austin (1793–1867), also responsible for translating Raumer's work and a dozen more German texts. The daughter of a Norwich wool-stapler, she was a friend of both Jeremy Bentham and James Mill. In the 1850s she edited the memoirs of Sydney Smith.

3 Raumer (1836), I, p. 174.

4 Letter of 27 October 1847, quoted by Feaver (1975), p. 144.

5 'Christopher North' in the Edinburgh Review, quoted by Johnstone (1974), p. 75. *The Deluge* is in a private collection.

6 Quoted by Pendered (1923), p. 180.

7 From *Arnold's Magazine of the Fine Arts 1833* quoted by Feaver (1975), p. 207.

8 Bulwer-Lytton (1833), pp. 211–12.

9 Bulwer-Lytton (1833), p. 218.

10 Turner painted at least three versions of the scene, of which one is a watercolour. The most celebrated of these is housed at the Philadelphia Museum of Art.

11 At the heart of the Irish problem were the imposition of Anglican tithes on a population overwhelmingly Catholic, and the policy of handing government jobs exclusively to its Protestant supporters in the province.

12 Quoted by Crowther (1965), p. 68.

13 Raumer (1836), I, p. 191.

14 Raumer (1836), I, pp. 192–3.

15 Raumer (1836), II, p. 134.

16 The Scot John Murray founded his small publishing firm in London in 1768. His son became the publisher of Jane Austen, Maria Rundell (who made his fortune with a cookery book), Benjamin Disraeli and Lord Byron. He was an accessory to the destruction of Byron's papers in 1824, which was opposed vehemently by John Martin's poet friend Thomas Moore, author of *Lallah Rookh*.

17 Bulwer-Lytton (1833), p. 116.

18 Henry Brougham, speech to Parliament, 29 January 1829, quoted in Brewer (1978).

19 Lane (1978), pp. 124–5.

20 Noble (1938), p. 116.

21 Noble (1938), p. 124.

22 Noble (1938), p. 138.

23 Noble (1938), p. 83.

24 Select Committee on Accidents in Mines, Chairman J. Pease, HMSO, September 1835.

25 From *Poems during a Summer Tour 1833*, XLII; Smith (1908), p. 320.

26 *Newcastle Weekly Chronicle*, 20 April 1889.

27 *Newcastle Weekly Chronicle*, 2 March 1889.

28 Acland (1948), p. 59.

29 *A Voice from the Factories*, X, 1836, London, John Murray.

CHAPTER FOURTEEN Survivors

1 Quoted by Mary Pendered (1923), p. 217.

2 Quoted by Balston (1945), pp. 101–2.

3 Quoted by Pendered (1923), p. 221.

4 Leopold Martin, *Newcastle Weekly Chronicle*, 16 March 1889.

5 Joseph Bramah, then the Brunels, and later the Martins themselves lived in a grand house on Lindsey Row (now Cheyne Walk), part of an ancient and grand complex that had been built in the seventeenth century. James McNeil Whistler would later live in part of the same buildings.

6 The old woman in question was Turner's lover Sophia Booth. This reference rather proves that no one was aware of their relationship until long after Turner's death.

7 Leopold Martin, *Newcastle Weekly Chronicle*, 16 March 1889.

8 Captain Collingwood (so far as one can tell no relation to Admiral Collingwood, 1748–1810) was commander of the slaveship *Zong* in 1781. Fearing that many of his 'cargo' would die

before reaching the West Indies, he ordered 132 of them to be thrown overboard so that he would not lose his insurance. The painting now belongs to the Boston Museum of Fine Arts.

9 Ruskin, in his 1840 essay 'The Proper Shape of Pictures' quoted by Balston (1947), p. 60.

10 Quoted by Balston (1947), p. 215. *The Stones of Venice*, the printed version of which excised these lines, was first published in 1851, by which time John Martin's reputation and fortune were immune to such criticism, as they had earlier been to his exclusion from the Royal Academy.

11 Quoted by Bowers (1975), p. 60.

12 Bowers (1975), p. 102.

13 Joseph Lancaster (1778–1836). His 1803 *Improvements in Education* was a reworking of the so-called Madras system, now referred to as peer tutoring, in which corporal punishment was replaced by moral instruction and older pupils were rewarded for teaching younger children. It was a system widely adopted before public education, and influential throughout the nineteenth century.

14 Hyman (1982), pp. 64–5.

15 Wymer (1958), p. 64.

16 Wymer (1958), p. 66.

17 Wymer (1958), p. 70.

18 Acland (1948), p. 100. The same letter is quoted by Lord David Cecil in his 1939–54 biography of Lord Melbourne.

19 Acland (1948), p. 100.

20 Dismissed from the Post Office in 1842 for party reasons, Hill became a director of the London to Brighton Railway. In 1844 he formed a society called the Friends in Council with Edwin Chadwick and John Stuart Mill, and also became a member of the Political Economy Club. In 1860 he became a Knight Commander of the Order of the Bath. He died in 1879 and is buried in Westminster Abbey.

21 Charles Dickens (1841), *The Old Curiosity Shop*, chapter 45.

CHAPTER FIFTEEN Judgement

1 The House of Lords sat in the ruins of the Painted Chamber, patched up to accommodate them temporarily after the fire and not finally pulled down until 1847. The House of Commons sat in the old Court of Requests, which was fitted up as a temporary chamber. This note has kindly been supplied by Ms Mari Takayanagi, Archivist, Parliamentary Archives, Houses of Parliament.

2 Karl Marx, *Dissertation and Preliminary Notes on the Difference between Democritus' and Epicu-*

rus' Philosophy of Nature, presented to the University of Jena in April 1841. Text quoted by McLellan (1971), p. 14.

3 Quoted by McLellan (1971), p. 13.

4 From Karl Marx's *Eighteenth Brumaire of Louis Bonaparte*.

5 From the online *Marx and Engels Collected Works*, vol. 1, pp. 374–5. The anonymous lithograph was also published to accompany the article 'The Grub Street Years', by Christopher Hitchens (the *Guardian*, Saturday 16 June 2007).

6 West and Walk (1977), p. 13.

7 13 September 1842, quoted by Acland (1948), p. 135.

8 *The Communist Party Manifesto*, English translation by Miss Helen Macfarlane (1850).

9 *Edinburgh Review*, June 1829.

10 *The Communist Party Manifesto*, English translation by Miss Helen Macfarlane (1850).

11 Babbage (1832), *On the Economy of Machinery and Manufactures*.

12 *The Communist Party Manifesto*, English translation by Miss Helen Macfarlane (1850).

13 McCartney and Arnold (1999), p. 5.

14 William Wordsworth, *Miscellaneous Sonnets*, XLV, Oxford, Clarendon Press.

15 Unknown banker to Charles Babbage, quoted by Morrison and Morrison (eds) (1961), p. 113.

16 Information from an unpublished article by Kevin Gordon, British Transport Police, 2003.

17 Information from an unpublished article by Kevin Gordon, British Transport Police, 2003.

18 Quoted by Bowers (1975), p. 139.

19 *Newcastle Weekly Chronicle*, 6 April 1889.

20 *Newcastle Weekly Chronicle*, 16 March 1889.

21 *Newcastle Weekly Chronicle*, 16 March 1889.

22 *Newcastle Weekly Chronicle*, 16 March 1889.

23 Quoted by Wilfred Blunt (1975).

24 Martin's legacy is persuasively examined in William Feaver (1975), *The Art of John Martin*, p. 208.

25 William Bell Scott (1811–90) is perhaps best known for painting a famous series of murals of northeast history at Wallington Hall, near Morpeth in Northumberland. A friend of the Rossettis, he lived and worked in Newcastle for 20 years after 1844. The murals survive in the Hall, which is owned by the National Trust.

26 William Bell Scott (1892), *Autobiographical Notes*, p. 196.

Bibliography

Ackroyd, P. (2006), *Turner*. London, Vintage.

Acland, A. (1948), *Caroline Norton*. London, Constable.

Adams, M. (2005), *Admiral Collingwood: Nelson's Own Hero*. London, Weidenfeld & Nicolson.

Adamson, W. (1877), *Notices of the Services of the 27th Northumberland Light Infantry Militia*. Newcastle, Robert Robinson.

Ainsworth, Harrison, W. (1841), *Old Saint Paul's*. London, Nelson edition.

Aldridge, A. O. (1960), *Man of Reason: The Life of Thomas Paine*. London, Cresset Press.

Alexander, B. (ed.) (1957), *Life at Fonthill 1807–22, from the Correspondence of William Beckford*. London, Rupert Hart-Davis.

Ayling, S. (1972), *George the Third*. London, William Collins.

Babbage, C. (1830), *Reflections on the Decline of Science in England*. London, B. Fellowes.

Babbage, C. (1832), *On the Economy of Machinery and Manufactures*. Philadelphia, Carey & Lea.

Balston, T. (1945), *The Life of Jonathan Martin*. London, Duckworth.

Balston, T. (1947), *John Martin*. London, Duckworth.

Bamford, S. (1844), *Passages in the Life of a Radical*. Fitzroy edition published by MacGibbon and Kee (1967).

Beckford, W. (1957), *Life at Fonthill*. London, Rupert Hart-Davis.

Bernal, J. D. (1969), *Science in History* 2: *The Scientific and Industrial Revolutions*. Harmondsworth, Penguin.

Black, J. and P. Woodfine (eds) (1988), *The British Navy and the Use of Naval Power in the Eighteenth Century*. Leicester, Leicester University Press.

Blackman, J. (1975), 'The Cattle Trade and Agrarian Change on the Eve of the Railway Age', *Agricultural History Review* 23:1, 48–62.

Blunt, W. (1975), *England's Michelangelo: A Portrait of George Frederick Watts*. London, Hamish Hamilton.

Bonser, K. J. (1970), *The Drovers*. London, Macmillan.

Bowers, B. (1975), *Sir Charles Wheatstone*. London, HMSO.

Boyle, J. R. (ed.) (1889), *The First Newcastle Directory 1778*. Newcastle, Mawson, Swan and Morgan.

Brewer, E. C. (1978), *Dictionary of Phrase and Fable*. New York, Avenel Books.

Briggs, A. (1979), *Iron Bridge to Crystal Palace: Impact and Images of the Industrial Revolution*. London, Thames & Hudson.

Brock, M. (1973), *The Great Reform Act*. London, Hutchinson.

Bulwer-Lytton, E. (1833), *England and the English*, 2 vols. London, Richard Bentley.

Burke, E. (1790), *Reflections on the Revolution in France*. Stanford University Press edition Stanford, (2001).

Campbell, W. (1853), *The Poetical Works of Thomas Campbell*. London, Edward Moxon.

Cecil, Lord D. (1954), *Melbourne*. New York, Grosset & Dunlap.

Chapman, S. D. (1972), *The Cotton Industry in the Industrial Revolution*. London, Macmillan.

Clark, K. (1971), *Civilisation*. London, BBC.

Clarke, J. (1977), *The Price of Progress: Cobbett's England 1770–1835*. London, Granada.

Clayre, A. (ed.) (1977), *Nature and Industrialisation*. Oxford, Oxford University Press.

Clubbe, J. (1997), 'Between Emperor and Exile: Byron and Napoleon 1814–1816', *Journal of the International Napoleonic Society*, 1:1.

Coad, J. (2005), *The Portsmouth Block Mills*. Swindon, English Heritage.

Coleman, T. (1968), *The Railway Navvies*. Harmondsworth, Pelican.

Collins, E. J. T. (1987), 'The Rationality of Surplus Agricultural Labour: Mechanisation in English Agriculture in the Nineteenth Century', *Agricultural History Review* 35:I, 33–46.

Cormack, P. (1981), *Westminster: Palace and Parliament*. London, Frederick Warne.

Crowther, J. G. (1965), *Statesmen of Science*. London, Cresset Press.

Derry, T. K. and T. I. Williams (1961), *A Short History of Technology*. New York, Dover.

Dinwiddy, J. R. (1986), *From Luddism to the First Reform Bill*. Historical Association Studies. Oxford, Basil Blackwell.

Donoghue, D. (1973), *Thieves of Fire*. London, Faber & Faber.

Fara, P. (2004), *Pandora's Breeches: Women, Science and Power in the Enlightenment*. London, Pimlico.

Feaver, W. (1975), *The Art of John Martin*. Oxford, Clarendon Press.

Finberg, A. J. (1961), *The Life of J. M. W. Turner RA*. Oxford, Clarendon Press.

Ford, B. (ed.) (1963), *From Blake to Byron: The Pelican Guide to English Literature* 5. Harmondsworth, Penguin.

Fraser, C. M. and K. Emsley (1973), *City and County Histories: Tyneside*. Newton Abbot, David & Charles.

Fraser, F. (1996), *The Unruly Queen: The Life of Queen Caroline*. London, Macmillan.

Frith, W. P. (1887), *My Autobiography and Reminiscences*. London, Richard Bentley.

Fynes, R. (1873), *The Miners of Northumberland and Durham*. 1985 reprint. Whitley Bay, Petrie.

Galloway, R. (1882), *A History of Mining in Great Britain*. London, Macmillan.

Gatt, G. (1968), *Constable*. London, Thames & Hudson.

Gaunt, W. (1964), *A Concise History of English Painting*. London, Thames & Hudson.

Gleick, J. (2003), *Isaac Newton*. London, Fourth Estate.

Godwin, W. (1793), *An Enquiry concerning Political Justice*. London, Joseph Johnson.

Godwin, W. (1794), *Caleb Williams*. London, B. Crosby.

Godwin, W. (1797), 'Reflections on Education, Manners and Literature', *The Enquirer*. London, Robinson.

Gore, J. (1934), *Creevey's Life and Times: A Further Selection from the Correspondence of Thomas Creevey*. London, John Murray.

Gottschalk, Louis R. (1927), *Jean-Paul Marat: A Study in Radicalism*. London, George Allen & Unwin.

Gray Perkins, J. (1909), *The Life of Mrs Norton*. London, John Murray.

Guy, A. (2003), *Steam and Speed: Railways of Tyne and Wear from the Earliest Days*. Newcastle upon Tyne, Tyne Bridge Publishing.

Halévy, E. (1960), *A History of the English People in the 19th Century*. Volume 1: *England in 1815*. London, Benn. 2nd edition.

Hall, S. C. (1883), *Retrospect of a Long Life*, 2 volumes. London, Richard Bentley.

Hamilton, J. (1998), *Turner and the Scientists*. London, Tate Gallery.

Hamilton, J. (2002), *Faraday: The Life*. London, HarperCollins.

Hardy, T. (1799), *An Account of the Origin of the London Corresponding Society*.

Harris, C. L. (1981), *Evolution Genesis and Revelations: With Readings from Empedocles to Wilson*. Albany, State University of New York.

Hartley, Sir H. (1966), *Humphry Davy*. London, Thomas Nelson & Sons.

Hazlitt, W. (1825), *The Spirit of the Age*. Collins Annotated Student Texts edition. London, 1969.

Hewitson, T. L. (1999), *A Soldier's Life: The Story of Newcastle Barracks*. Newcastle, Tyne Bridge.

Hibbert, C. (1969), *London: The Biography of a City*. London, Longman Green.

Hobsbawm, E. (1962), *The Age of Revolution*. London, Abacus. Paperback edition.

Hobsbawm, E. and T. Ranger (eds) (1983), *The Invention of Tradition*. Cambridge, Cambridge University Press.

Horsley, P. M. (1971), *Eighteenth Century Newcastle*. Newcastle, Oriel Press.

Hunt, J. H. L. (attrib) (1825), *Rebellion of the Beasts*. Chicago, Wicker Park Press edition (2004).

Hunt, J. H. L. (1850), *Autobiography of Leigh Hunt*. London.

Hutchinson, T. (ed.) (1929), *The Complete Poetical Works of Percy Bysshe Shelley*. London, Oxford University Press.

Hyman, A. (1982), *Charles Babbage: Pioneer of the Computer*. Oxford, Oxford University Press.

Jennings, H. (1985), *Pandaemonium: The Coming of the Machine as seen by Contemporary Observers*. London, André Deutsch.

Johnstone, C. (1974), *John Martin*. London, Academy Editions.

Jones, W. L. (1983), *Ministering to Minds Diseased: A History of Psychiatric Treatment*. London, Heinemann Medical.

Kegan Paul, C. (1876), *William Godwin: His Friends and Contemporaries*. London, Henry S. King.

Klingender, F. D. (1972), *Art and the Industrial Revolution*. St Albans, Paladin.

Knight, D. (1992), *Science and Power*. Oxford, Blackwell.

Lane, P. (1978), *Success in British History*. London, John Murray.

Loomis, S. (1964), *Paris in the Terror*. London, Penguin.

Maas, J. (1969), *Victorian Painters*. London, Barrie & Jenkins.

McCartney, S. and A. J. Arnold (1999), *A Very Bold and not at all Unwise Projector: George Hudson at the York and North Midland Railway 1833–49*. Colchester, University of Essex.

Macdonald, S. (1975), 'The Progress of the Early Threshing Machine', *Agricultural History Review* 23:1, 63–77.

Macdonald, S. (1979), 'The Diffusion of Knowledge among Northumberland Farmers 1780–1815', *Agricultural History Review* 27:I, 30–9.

Macfarlane, C. (1917), *Reminiscences of a Literary Life*. London, John Murray.

McLellan, D. (1971), *Karl Marx: Early Texts*. Oxford, Basil Blackwell.

Marat, J.-P. (1774), *The Chains of Slavery*. Newcastle.

Martin, J. (1829), *The Life of Jonathan Martin the Incendiary of York Minster*. Leeds, Robinson, Hernaman & Woods: 6th edition.

Martin, T. (1934), *Faraday*. London, Duckworth.

Martin, W. (1832), *The Defeat of Learned Humbugs and the Downfall of All False Philosophers*. Wallsend.

Martin, W. (1833), *A Short Outline of the Philosopher's Life from being a Child in Frocks to the Present Day after the Defeat of all Imposters, False Philosophers, since the Creation*. Newcastle upon Tyne.

Marx, K. and Engels, F. (1848), *The Communist Manifesto*. London, Penguin Great Ideas edition (2004).

Mathias, P. (1983), *The First Industrial Nation*. London, Methuen, 2nd edition.

Maxwell, Sir H. (1905), *The Creevey Papers*. London, John Murray, 3rd edition.

Middleton, C. (ed.) (1983), *Goethe, Selected Poems*. Princeton, Princeton University Press.

Minchinton, W. E. (1953), 'Agricultural Returns and the Government during the Napoleonic Wars', *Agricultural History Review* 1, 29–43.

Monkhouse, C. (1893), *The Life of Leigh Hunt*. Walter Scott.

Morrison, P. and E. Morrison (eds) (1961), *Charles Babbage and his Calculating Machines*. New York, Dover.

Myrone, M. (2006), *Gothic Nightmares: Fuseli, Blake and the Romantic Imagination*. London, Tate Gallery.

New Lanark Conservation Trust (2002), *The Story of New Lanark*. New Lanark.

Nicholson, W. E. (1888), *Glossary of Terms used in the Newcastle Coalfield*. Newcastle upon Tyne.

Noble, C. B. (1938), *The Brunels*. London, Cobden-Sanderson.

Norton, C. (1830), *The Undying One*. London, Henry Colburn.

Owen, R. (1813), *A New View of Society*. Everyman edition: London, John Dent (1927).

Padfield, P. (2003), *Maritime Power*. London, John Murray.

Paine. T. (1776), *Common Sense*. Philadelphia, W & T Bradford.

Peacock, T. L. (1831), *Crotchet Castle*. London, T. Hookham.

Pearce Williams, L. (1965), *Michael Faraday*. London, Chapman & Hall.

Pendered, M. L. (1923), *John Martin, Painter*. London, Hurst & Blackett.

Phillips, Sir R. (1820), *A Morning's Walk from London to Kew*. 2nd edition, London.

Pope, W. B. (ed.) (1963), *The Diary of Benjamin Robert Haydon*. Cambridge, MA, Harvard University Press.

Pückler-Muskau, Prince (1832), *Tour in England, Ireland and France in the Years 1828 and 1829*. Translated by Sarah Austin. London, Effingham Wilson.

Raumer, F. von (1836), *England in 1835*. 3 volumes. Translated by Sarah Austin. London, John Murray.

Richardson, B. W. (1891), *Thomas Sopwith*. London, Longman.

Richardson, J. (1961), *My Dearest Uncle: Leopold I of the Belgians*. London, Jonathan Cape.

Roe, N. (2005), *Fiery Heart: The First Life of Leigh Hunt*. London, Pimlico.

Rogers, S. (1859), *Recollections*. London, Longman.

Rolt, L. T. C. (1960), *George and Robert Stephenson: The Railway Revolution*. London, Longman.

Rousseau, J.-J. (1750), *Discourse on the Arts and Sciences*. Everyman edition: London, John Dent (1913).

Royle, E. (1986), *Chartism*. Seminar Studies in History. London, Longman, 2nd edition.

Ruskin, J. (1843–60), *Modern Painters: A Volume of Selections*. London, Nelson.

Scott, W. B. (1892), *Autobiographical Notes*. London, Osgood, McIlvaine.

Second J. A. (2000), *Victorian Sensation: The Extraordinary Publication, Reception, and Secret Authorship of Vestiges of the Natural History of Creation*. Chicago, University of Chicago Press.

Seznec, J. (1964), *John Martin en France*. London, Faber & Faber.

Shelley, M. (1818), *Frankenstein, or, the Modern Prometheus*. London, Wordsworth Classic edition (1993).

Shelley, M. (1826), *The Last Man*. London, Henry Colburn. Oxford, World's Classics edition (1994).

Singer, C. J. (ed.) (1958), *A History of Technology*. Oxford, Oxford University Press.

Smiles, S. (1863), *Industrial Biography: Iron Workers and Tool Makers*. London, John Murray.

Smith, A. (1776), *An Inquiry into the Nature and Causes of the Wealth of Nations*. London, W. Strahan & T. Cadell.

Smith, E. A. (1999), *George IV*. London, Yale University Press.

Smith, N. C. (ed.) (1908), *The Poems of William Wordsworth*. London, Methuen.

Sykes, J. (1833), *Local Records, or Historical Register of Remarkable Events*. Facsimile edition of 1973. Newcastle, Patrick & Horton.

Taylor, N. and A. Hankinson (1996), *Twelve Miles from a Lemon: Selected Writings and Sayings of Sydney Smith*. Cambridge, Lutterworth Press.

Thackeray, W. M. (1855), *The Four Georges*. Pocket Classics edition. London, Alan Sutton (1995).

Thompson, R. (2005), *Thunder Underground: Northumberland Mining Disasters 1815–65*. Newcastle, Landmark.

Todd, R. (1946), *Tracks in the Snow: Studies in English Science and Art*. London, Grey Walls.

Tomalin, C. (1980), *Shelley and His Times*. London, Thames & Hudson.

Treasure, G. (1992), *Who's Who in Early Hanoverian Britain*. London, Shepheard-Walwyn.

Treasure, G. (1997), *Who's Who in Late Hanoverian Britain*. London, Shepheard-Walwyn.

Trelawny, E. J. (1878), *Memoirs of Shelly, Byron and the Author*. London, Basil Montagu Pickering.

Viztelly, H. (1893), *Glances back through Seventy Years*. London, Kegan Paul.

Ward, H. (1926), *History of the Athenaeum*. London, Athenaeum Club.

Ward, J. Y. (ed.) (1970), *Popular Movements c. 1830–1850*. London, Macmillan.

Watson, W. F. (1998–9), 'The Invention of the Miners' Safety Lamp: A Reappraisal', *Transactions of the Newcomen Society* 70, 135–41.

Webb, J. (1827), *The Mummy*. London, Henry Colburn. Abridged edition, Ann Arbor Paperbacks (1994).

Wells, R. (1997), 'Mr. William Cobbett, Captain Swing, and King William IV', *Agricultural History Review* 45, 34–48.

West, D. J. and A. Walk (1977), *Daniel McNaughton: His Trial and the Aftermath*. Gaskell.

Williams, M. (1971), *Revolutions 1775–1830*. Harmondsworth, Penguin.

Wollstonecraft, M. (1792), *A Vindication of the Rights of Woman*. London, Joseph Johnson.

Woodcock, G. (1946), *William Godwin*. London, Porcupine Press.

Wordsworth, W. (1858), *The Poetical Works of William Wordsworth*. London, Routledge.

Wymer, N. (1958), *Social Reformers*. Oxford, Oxford University Press.

Ziegler, P. (1973), *King William IV*. London, Fontana.

Index